# ORIGAMI TESSELLATIONS
## Awe-Inspiring Geometric Designs

# ORIGAMI TESSELLATIONS

## Awe-Inspiring Geometric Designs

by Eric Gjerde

A K Peters, Ltd.
Wellesley, Massachusetts

Editorial, Sales, and Customer Service Office
A K Peters, Ltd.
888 Worcester Street, Suite 230
Wellesley, Massachusetts 02482
www.akpeters.com

Cover design by Jeffrey Rutzky
Production and editing by Jeffrey Rutzky
Back cover: *Star Burst* by Alex Bateman; *Mask 1A* by Joel Cooper

Library of Congress Cataloging-in-Publication Data
Gjerde, Eric, 1978–
  Origami tessellations : awe-inspiring geometric designs /
Eric Gjerde. — 1st ed.
    p. cm.
    Includes index.
    ISBN 978-1-56881-451-3 (alk. paper)
1. Origami.  I. Title.
    TT870.G49 2008
    736'.982—dc22

                                        2007020769

Printed in Canada

12  11  10  09  08    10 9 8 7 6 5 4 3 2 1

# Contents

# Preface

When I was five, my parents asked me what I wanted to be when I grew up. "A paperologist," I replied enthusiastically.

Throughout my early childhood and teen years, I enjoyed paper crafts and origami. I found imagining a thing and then creating it from materials at hand very satisfying. Friends and family frequently gave me stacks of papers and rolls of tape as birthday gifts.

Paperology wasn't a major at my college of choice, however. So, after preparing for a technology career, I found I still needed an artistic outlet. I turned back to my childhood love of paper and origami.

How was I drawn to the geometric art of origami tessellations, specifically? It started with a fish.

I was folding an origami koi designed by Robert J. Lang. The fish-scale design was folded from repeated patterns called tessellations. I practiced a bit on some scraps of paper, and, just like a fish, I was hooked on tessellations. I folded sheet after sheet of paper, experimenting with different shapes and variations. My excitement led to deeper exploration and the creation of new tessellation designs. I never looked back; in fact, I never did complete that fish!

Since then I have folded hundreds of origami tessellations. Tessellations are geometric patterns that can repeat forever. When captured in paper, beautiful pieces of art are created in intricate pleats and folds.

For this book, I have taken 25 of my favorite tessellations and turned them into projects for newcomers and experienced origamists alike. With step-by-step instructions, illustrated crease patterns, and how-to photos, you'll learn to create these wonderful designs yourself—hopefully, more easily than I did!

This book covers the fundamentals of origami tessellations, providing a little history and describing simple beginning techniques with detailed illustrations. The techniques then are utilized in specific projects that will help you make impressive works of art. Finally, the book's gallery of tessellation images is designed to inspire you to experiment and innovate, trying out the tessellations of others and, eventually, creating your own patterns.

## NO CUTTING, NO GLUING, ONE SHEET OF PAPER

All you need to make origami tessellations are your two hands and a piece of paper. Folding tessellations is not complicated. At its core, it is the repetition of simple folding patterns to create larger, more complex designs.

Folding tessellations is a meditative experience. While origami is labor-intensive, it's also very relaxing. Often when I fold something I lose track of time. Hours pass without my knowledge. When I'm finished, the completed piece is almost like a present.

When I teach students how to create tessellations, they always have an "Aha!" moment. They suddenly, unexpectedly understand how the folded shapes repeat themselves and how they all work together. From that point on the students take off and start folding madly. Three words of warning: Tessellations are addictive!

Don't worry if your first project comes out imperfectly; your next attempt will be greatly improved. Your own "Aha!" moment will come soon. Practice and repetition are great teachers.

As your skill and experience increase, you'll start to see tessellation patterns everywhere. Tiles on the floor, cobblestones in the courtyard, the patterns of light falling through a stained-glass window—you'll find inspiration all around you.

You may find discovery and creation are the most rewarding parts of this craft, because making origami tessellations is very much about infinite possibility. When you draw upon your newfound recognition of patterns and think about how they might be recreated in paper, you can use the techniques in this book to create amazing arrangements and artwork all your own.

# INTRODUCTION AND TECHNIQUES

# Overview

## WHAT IS A TESSELLATION?

**H**ave you ever looked at the patterns on a tile floor? The tiles you saw likely created tessellations—repeating patterns of specific shapes. The word "tessellation" comes from the Latin *tessella*, meaning "small square." The Romans frequently used squares for making mosaics and tile designs.

While the Romans and Byzantines made complex mosaic patterns, tessellations were raised to a true art form by artisans of the Islamic faith. Since Islam forbade representational imagery, artists and craftsmen focused their creativity on developing complex geometric designs. Their geometric orientation was reflected in inspiring architecture, tile work, painting, ceramics, and illustration.

The Islamic Moors in Spain built the Alhambra, a magnificent palace decorated with some of the best tessellation artwork in the world. The Dutch artist M. C. Escher (1898–1972) made several visits to the Alhambra. The complex tessellation patterns he studied had a formative impact on his art. Escher said the tiling at the Alhambra was "the richest source of inspiration I have ever tapped."

Now Escher's own art is among the most widely known and loved in the world. Tessellations are an important part of many of his sketches and illustrations, which feature fantastic animals, figures, and shapes.

## ORIGAMI TESSELLATIONS

Origami tessellations are geometric designs folded from a single sheet of paper, creating a repeating pattern of shapes from folded pleats and twists. Tessellated origami pieces range from simple square tilings to extremely intricate, complex pieces inspired by Islamic art. Tessellated shapes can form everything from twisted architectural flourishes to realistic faces.

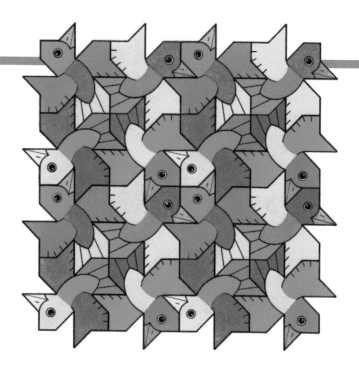

*Tessellating Birds* by David Bailey

Today's origami tessellations are primarily the brainchild of Shuzo Fujimoto, a Japanese chemistry teacher who was instrumental in exploring the possibilities of folding repeating shapes on paper. Fujimoto's singular focus on origami tessellations continues to inspire many folders. Most of the basic tessellation ideas covered in this book were discovered by Fujimoto-sensei in the 1970s.

Many people have contributed to this art form's development, including the computer scientist Ron Resch, prolific origami author Yoshihide Momotani, and artist Chris Palmer. Not long after spending six months living in the caves of Sacremonte outside of Granada, Spain, on an artistic journey to study the tilings of the Alhambra, Palmer was exposed to the tessellations of Fujimoto.

Building on Fujimoto's work, Palmer has pushed the art of origami tessellations upward and outward. He is one of the greatest tessellation folders today, and many tessellation artists count Palmer among their primary influences.

# BASIC TILINGS

Three very basic tessellation patterns, called "regular tessellations," are used heavily in origami tessellation designs. The three tiling patterns are formed with single, repeating shapes: equilateral triangles, squares, and hexagons. Often these patterns are referred to as the 3.3.3.3.3.3, 4.4.4.4, and 6.6.6 tessellations, respectively.

Separating numbers by periods is a common way of describing regular and semiregular tessellations by the number of corners of the shapes surrounding a given single intersecting point, or vertex. Because the regular tessellations are created from a single shape, their naming is relatively straightforward. 3.3.3.3.3.3 explains that there

are six shapes with three corners surrounding any given interior point.

In semiregular tessellations, the types of shapes differ, but patterns still are based on the arrangement of shapes around a single point. For any given point in the tessellation, the same shapes must be present and arranged in the same order. Three very common examples used in origami tessellations are the 3.6.3.6, 3.4.6.4, and 8.8.4 tessellations.

Origami tessellations often follow one of these six tessellation geometries by employing a sheet of paper precreased with a geometric grid.

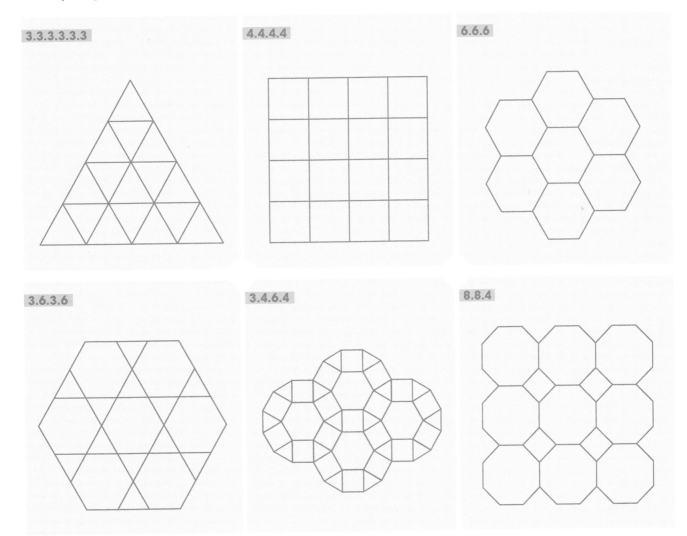

# MATERIALS

Origami tessellations require very little in the way of materials or tools—only a sheet of paper and your hands are needed. So, the starter kit is pretty basic!

Not all papers are created equal, however. Because origami tessellations are made from many creases and folds, the paper you choose needs to be capable of handling all this activity without cracking or falling apart.

The best paper meets three criteria:
1. The paper is flexible enough to work easily.
2. The paper holds a crisp crease line.
3. The paper does not crack or tear when creased repeatedly.

Many regular wood-pulp papers fall short of the mark. They are prone to falling apart, because wood-pulp papers are made with short fibers. Although you still can create fantastic art with wood-pulp paper, choosing a better quality paper will make a big difference in the final results of your folding, as well as the overall ease of your work.

A variety of papers will serve you well. The most important thing is to find a paper you're comfortable folding and that suits your folding style as it develops. Try out different kinds of paper, and discover which one works best for you.

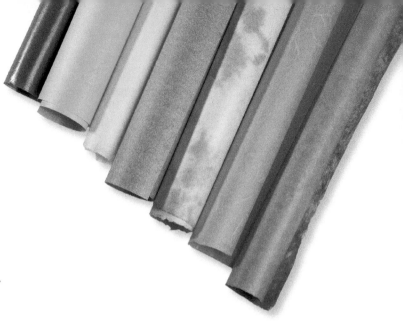

## Mulberry

A large selection of papers is made from the fibrous bark of the mulberry tree. Mulberry fibers are long, which lends great strength to the paper and also allows it to be made in very thin sheets. Mulberry paper includes varieties with names like washi, unryu, and saa.

All the mulberry papers require some preparation before folding, because they are soft and flexible. Lay the paper on a flat surface, such as a glass table, and spray it with a heavy spray starch normally used for ironing dress shirts. Press the paper flat and work out any bubbles. Let it dry. This treatment will create a stiff, crisp paper that still folds easily.

## Lokta

Made from the inner bark of the Daphne plant of Nepal, lokta paper has a soft, silken texture that makes it a pleasure to fold. Lokta has a natural ivory color, but often it is dyed with subtle or wild patterns.

Lokta also is available in a stiffer oiled variety that is particularly well-suited for tessellation folding.

## Elephant Hide

Made by the company M-real Zanders GmbH, Elephant Hide *(Elefantenhaut)* is a bookbinding paper widely used for origami tessellations.

Elephant Hide is heavier than many other origami papers. As a result, it is well-suited for three-dimensional pieces and also extremely resistant to tearing.

## Glassine

Glassine is very thin, translucent paper made from paper meal and some artificial additives. This paper is somewhat notorious for being difficult to use, because it has a tendency to crinkle easily.

However, no other paper can create the amazing displays that glassine provides when a finished origami tessellation is backlit. For this reason, some tessellation artists use glassine exclusively.

# Techniques

Learning a few basic techniques will launch you on your way to creating origami tessellation masterworks. Precreasing paper with square or triangle grids is a vital first step for creating many tessellations. Making pleat intersections (including inverted ones!) of varying degrees is another important, fundamental technique. Finally, learning the folds to make various "twists"—including triangle, square, and hexagon twists—will give you a rich inventory of origami skills. Almost every tessellation is constructed from a combination of these techniques.

As with the instructions for the 25 tessellation projects in this book, to some extent words on paper can take you only so far: Your *hands* on paper will be needed for you to understand the intent and meaning of some of the instructions. Once your hands are engaged, many "Aha!" moments are sure to follow!

## PRECREASING

One of the fundamental underpinnings for creating origami tessellations is the use of precreased grids. Typically, a grid is used as the framework for the placement and orientation of shapes and pleats, which match up with the geometry of the grid itself.

The two basic grid patterns are made up of squares and equilateral triangles, because all three regular tessellations (triangles, squares, and hexagons) can be created from these two grids.

Accuracy is very important when you fold the patterns. Because you are creating the majority of the creases in your tessellation when you make the grid, the pleats on your grid must be of equal width and all your lines must be parallel.

If you find you are a bit off at first, don't be surprised or discouraged. You'll improve with practice. Focus your efforts on making your first few folds as accurate as possible, because they're the cornerstone of the foundation of your grid.

Much traditional Islamic architecture, including the beautiful imagery featured throughout this chapter, is decorated with tessellating designs.

## Square Grid

**1** To fold a grid of squares, fold a square piece of paper in half (figure 1). Rotate the paper a quarter turn and fold again (figure 2). Fold each of these halves in half again, repeating this action in each direction (figure 3).

**2** Repeating the process again gives you a division of eight pleats in each direction (figure 4), and doing so a fourth time gives you divisions of 16 (figure 5).

You can repeat the process until you have enough pleats for the pattern you intend to fold. Quite often you will need to have divisions of 32 or even 64 pleats for large pieces.

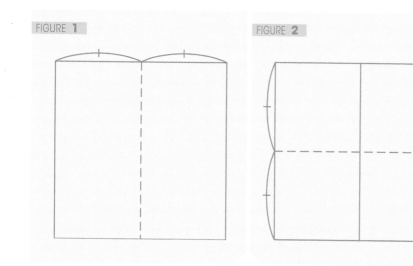

FIGURE **1**

FIGURE **2**

FIGURE **3**

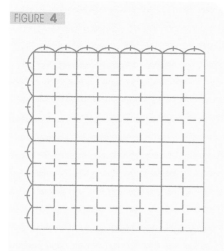

FIGURE **4**

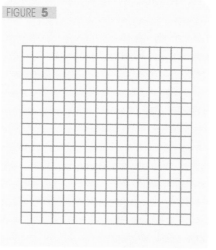

FIGURE **5**

## Triangle Grid

The equilateral triangle grid is the most commonly used grid pattern in modern tessellation designs, because both triangle and hexagon designs can be folded from it.

You can make this grid in two ways: The most common method is to fold the grid from reference lines on a square sheet of paper, which is the technique described in this section. The second method is to cut a hexagon from a sheet of paper and fold it in halves using the same folds as the ones used to create a square grid. Folding a hexagon actually is easier than folding a triangle grid on square paper, but cutting perfectly accurate hexagons takes a lot of practice. If you try this method, I recommend purchasing template guides for cutting hexagons, which are available from most craft stores.

**1** To fold a triangle grid from a sheet of square paper, fold the sheet in half vertically, taking care to crease in only about ½ inch (1 cm) from the edges of the paper (figure 6).

**2** Fold one side in half again, bringing a side of the sheet to the crease marks in the center; this time, pinch a crease about 2 inches (5 cm) long in the center to mark the quarter division (figure 7). Bring one of the opposite corners to this new crease line, pivoting the paper on the center-edge crease (figure 8).

**3** Repeat step 2 for the other side of the paper (figure 9). Rotate the paper 180 degrees and repeat the same process for the two remaining corners (figures 10 and 11).

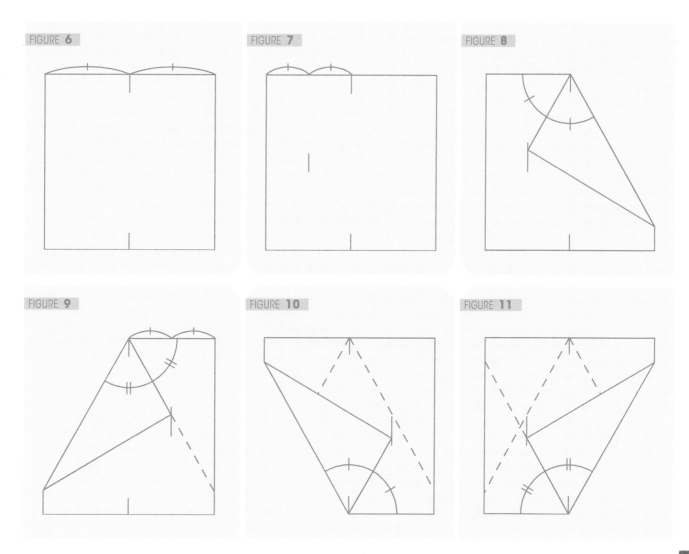

FIGURE **6**

FIGURE **7**

FIGURE **8**

FIGURE **9**

FIGURE **10**

FIGURE **11**

**4** Fold the paper in half horizontally, making a crease line that runs through the two existing crease intersections (figure 12). You now have created all the reference lines you will need to finish folding a grid of triangles!

**5** Fold the paper in quarters horizontally, based on the center crease you just made (figure 13). Now fold the diagonal creases together in each direction, creating parallel pleats halfway between them (figure 14). To fold the creases positioned toward the corners, carefully line up the paper with existing creases, and fold new creases to match them.

**6** Repeat the subdivision process until you have a triangle grid folded to your liking (figure 15). Triangle grids with 32 pleat divisions are commonly used in making origami tessellations, but 16-division grids are an easy way to get started. Very complex designs often require 64 pleat divisions or even more!

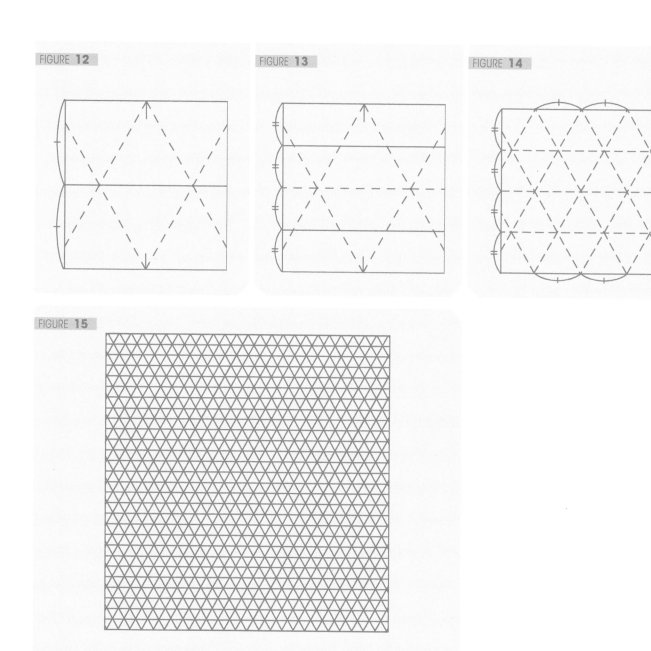

FIGURE **12**

FIGURE **13**

FIGURE **14**

FIGURE **15**

# PLEAT INTERSECTIONS

Learning to make pleat intersections is a bit like learning to walk. You need to practice (and maybe stumble at first) in order to get anywhere! And, just like with walking, the more you practice, the better you'll get at it.

Please note: Some of the illustrations for the techniques—pleat intersections and then twists—that follow are shown on hexagonal-shaped triangle grids or some variation for ease of explaining the maneuvers, but the folds would be the same on a square-shaped triangle grid. In fact, all these folds can be done in the middle of a grid that extends to infinity!

## 120-Degree Pleat Intersection

**1** Using a triangle grid, identify the three pleats you want to fold, and pinch the paper together to form the pleats (figure 16).

**2** When all the pleats are pinched together, fold the pleats over in the direction you want them to go (figure 17). Lay the pleats flat (figure 18).

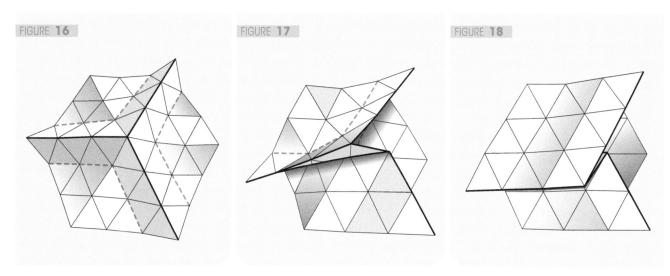

FIGURE **16**

FIGURE **17**

FIGURE **18**

# 120-Degree Inverted Pleat Intersection

Almost all pleats can be inverted. Inversion means a pleat lies on the back side of the design, rather than on the front of the surface. Pleat inversion is useful and practical, particularly when you are trying to reduce clutter in a busy design or maintain lines of symmetry for visual effect.

To invert a 120-degree pleat:

**1** Using a triangle grid, identify the three pleats you want to fold, and pinch the paper together to form the pleats (figure 19).

**2** Take one of the pleats and invert it by changing the mountain fold directed toward you into a valley fold directed away from you (figure 20). The two remaining mountain pleats now will naturally fold down in the direction of the inverted pleat. Please note that one triangle of the inverted pleat located at the intersection must remain a mountain fold (figure 21).

**3** Pinch together the inverted pleat and fold it flat on the reverse side of the paper. The finished pleat intersection will lie flat with all three of the pleat lines meeting up at one point (figure 22).

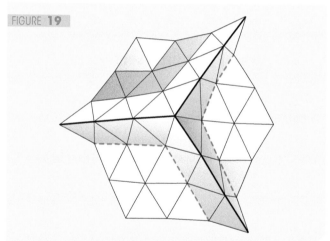
FIGURE **19**

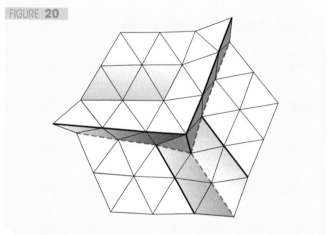
FIGURE **20**

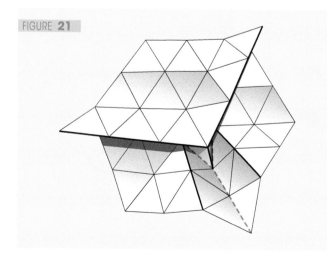
FIGURE **21**

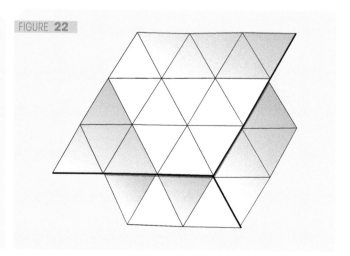
FIGURE **22**

## 90-Degree Pleat Intersection

**1** Using a square grid, fold a single pleat (figure 23). Unfold the pleat, and fold a second pleat crossing over the first one (figure 24).

**2** Unfold the second pleat, and then pinch the folds together along the diagonal creases shown in figure 25.

**3** Fold the lower flaps outward, and fold the tip over on the diagonal lines (figure 26). The paper should lie flat when you are finished (figure 27).

The 90-degree pleat intersection can be adapted in several ways. One common adaptation is folding the lower flaps inward. To fold this adaptation, change the orientation for the lower flaps so they fold inward rather than outward, and reverse the diagonal creases from valley to mountain folds (figure 28). The finished fold will lie flat on the paper and look a bit like a bird's mouth (figure 29).

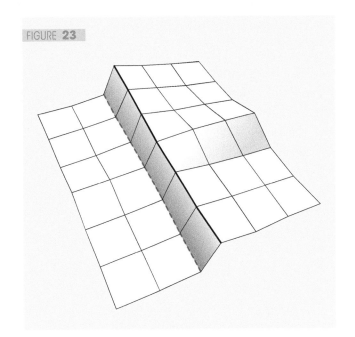
FIGURE 23

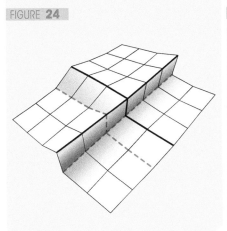
FIGURE 24

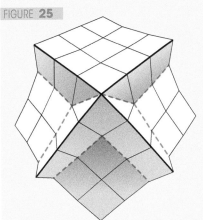
FIGURE 25

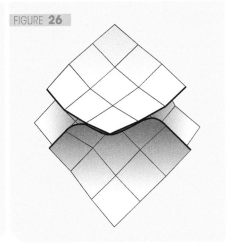
FIGURE 26

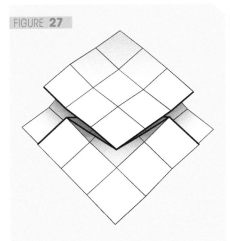
FIGURE 27

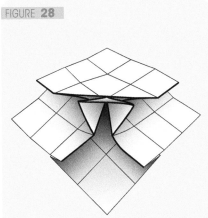
FIGURE 28

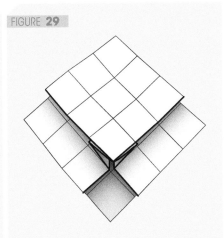
FIGURE 29

# 60-Degree Pleat Intersection

In the 60-degree pleat intersection, another very common fold, two pleats of the triangle grid cross. Folding these pleats so they continue in a straight line requires a little re-arrangement, because each pleat tries to displace the other. With a bit of modification, the lines flow undisrupted.

**1** Start by making a single pleat on a triangle grid (figure 30). Unfold the pleat, and add a second pleat crossing over the first one (figure 31).

**2** Unfold again, so you can see the triangular tip of the pleat intersection clearly. Change the pleats extending from the tip from mountain folds to valley folds (figure 32).

**3** Fold in the outer sides of the valley folds, collapsing the pleats on both sides of the triangular tip (figure 33). This move requires some adjustment of the paper to make it come together.

**4** Turn the entire piece over. Now looking at the back side, fold the two pleats over toward each other (figure 34). You will need to overlap the layers for the paper to lie flat (figure 35).

**5** Turn the piece back to the front side. The pleat intersection will come together and be locked into place by the overlapping layers (figures 36 and 37).

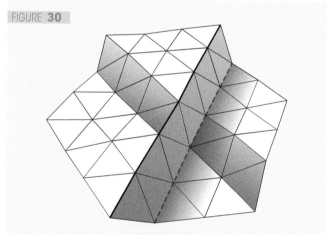

FIGURE 30

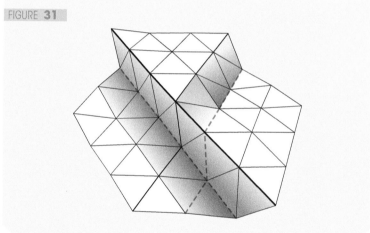

FIGURE 31

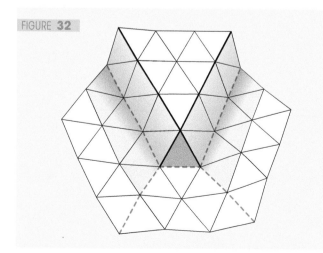

FIGURE 32

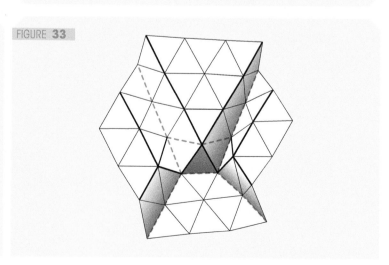

FIGURE 33

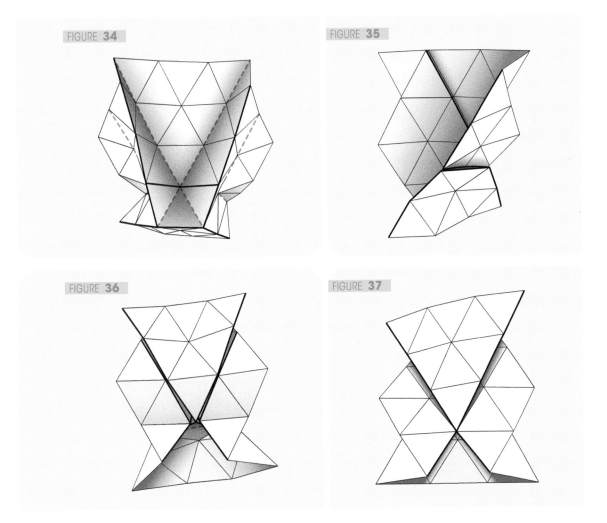

FIGURE **34**

FIGURE **35**

FIGURE **36**

FIGURE **37**

## Rabbit-Ear Triangle Sink

This pleat intersection is a somewhat tricky folding maneuver, but the terrific results are worth the effort involved. The rabbit-ear triangle sink basically is a 60-degree pleat intersection. However, instead of folding the pleats over each other, you create a rabbit-ear triangle sink by allowing all the pleats to lie in the same orientation on the surface of the paper.

You create pointed tips with this technique, and you can use these tips to add interesting effects and details to tessellations. This pleat intersection often is used to increase visual complexity and variety in star-shaped patterns and designs.

FIGURE **38**

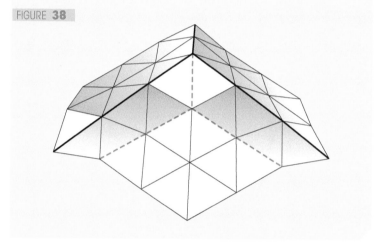

FIGURE **39**

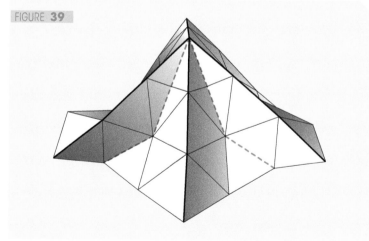

**1** On a triangle grid, create a 120-degree intersection. Fold two of the pleats toward you, creating two valley creases that meet together. Unfold (figure 38).

**2** Pull the center crease between the two valley creases outward, creating a mountain fold, and pinch two offset creases on either side of it, dividing the entire piece into 30-degree angles (figure 39).

**3** Fold another offset crease connecting to the other two, so a triangle is formed. Pinch this triangle together, so the tip of the intersection sticks upward and the pleats lie flat on the surface of the paper (figure 40).

**4** Fold the tip over in the direction you wish it to go and flatten the paper (figure 41). Turn the piece over. Looking at the paper from the reverse side, you can see how the triangle sink allows the pleat intersection to avoid overlapping layers, while also enhancing symmetry (figure 42).

**5** To complete the second half of the rabbit-ear triangle sink, turn the piece back to the front side's original positioning and rotate it 180 degrees. Repeat steps 1 to 4 on the other half of the paper. When you are finished, you will have created a freely folding triangular flap (figure 43).

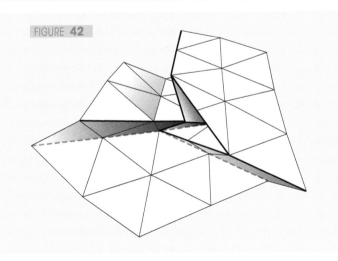

FIGURE **40**

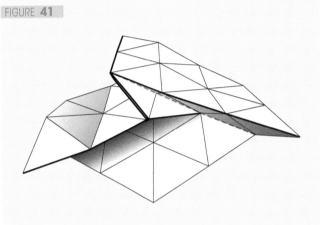

FIGURE **41**

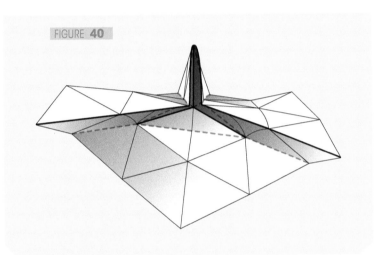

FIGURE **42**

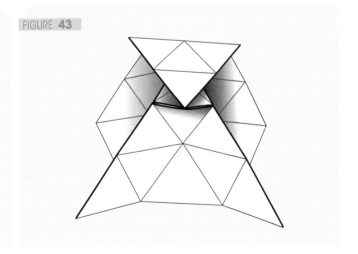

FIGURE **43**

# TWISTS

## Triangle Twist

Triangle twists are created when three pleats meet together. The most common type of triangle twist is made from three pleats of equal width.

Using a precreased grid of triangles makes folding triangle twists one of the easiest tessellation techniques. You can utilize triangle twists in a nearly unlimited number of designs—the only possible limits reside in your imagination and the amount of paper you have on hand!

**1** Start with a precreased grid of triangles. Fold three mountain pleats at 60-degree intervals (figure 44).

**2** When you pull the pleats in, you'll notice paper building up in the center. This extra paper is required for the paper to twist. Help the process along by folding the pleats over in the same direction (clockwise or counterclockwise), rotating around the central meeting point of the pleats. You will form a triangular peak at the center (figure 45).

**3** Hold down the three corners of the triangle, and pull on them slightly. You are exerting pressure on the center of the triangle and making it start to spread out. Encourage the process by pushing down on the center of the triangle (figure 46).

**4** Flatten this middle triangle section all the way to its corners (figure 47). Your triangle twist is complete!

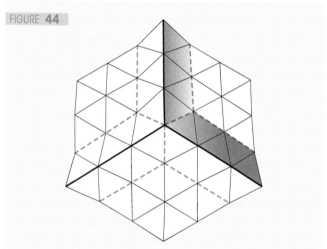

FIGURE **44**

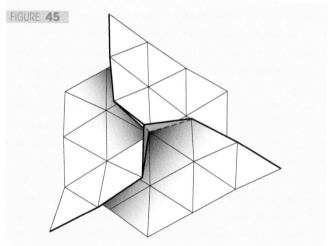

FIGURE **45**

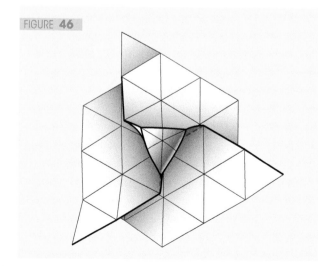

FIGURE **46**

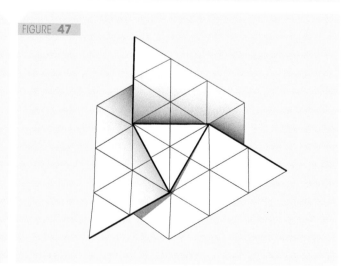

FIGURE **47**

## Square Twist

A square twist is the crossed intersection of two pleats running perpendicular to one another. Typically, the pleats have equal width. The square twist is easy to fold and very common in origami tessellations.

**1** Start with a precreased grid of squares. Identify the mountains and valleys of the two perpendicular pleats you want to fold, and precrease the diagonal creases (figure 48).

**2** Start folding the paper along these diagonal creases (figure 49), while folding the paper in half (figures 50 and 51).

**3** Fold the paper along the remaining unfolded valley folds, opening the paper up so it lies flat (figures 52 and 53). That's all there is to it!

Although in these instructions I've recommended you precrease the diagonal angles, after some practice you'll be able to squash the paper into place when doing your folding. Many tessellation folders find this squashing highly rewarding, almost like popping bubble wrap!

Also, with additional practice you'll be able to fold this twist without needing to fold the entire piece in half. Try folding some square twists by pinching together the four intersecting pleat segments and twisting them in a fashion similar to the triangle-twist method.

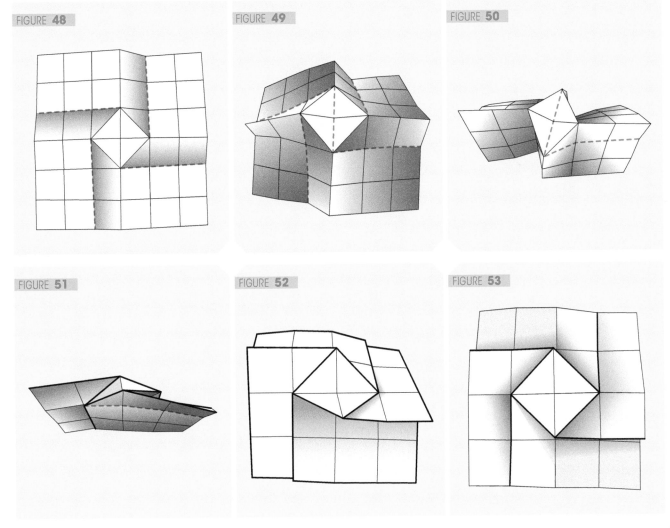

FIGURE **48**

FIGURE **49**

FIGURE **50**

FIGURE **51**

FIGURE **52**

FIGURE **53**

## Hexagon Twist

The hexagon twist, or hex twist, is made from six intersecting pleats on a triangle grid. The hex twist is a mainstay of origami tessellations. You'll often see hexagon shapes appear while folding, sometimes quite unexpectedly. Many tessellation patterns include at least one hexagon twist.

**1** On a precreased triangle grid, identify six intersecting pleats to fold with mountain and valley folds (figure 54). Fold the paper around the central hexagon (figure 55), working the paper so the hexagon closes itself in half, like a book (figures 56 and 57). The paper should collapse along the valley folds so the entire piece is folded in half (figure 58).

**2** With the twist now folded in half, open the two halves of the "book" while holding the base together (figure 59). The paper should open up and lie flat (figure 60).

Once the hex twist has been folded, it locks the paper into place, so it can be quite difficult to unfold. The hex twist is a very useful tool when you are folding complicated patterns, because it does a good job holding complex designs together.

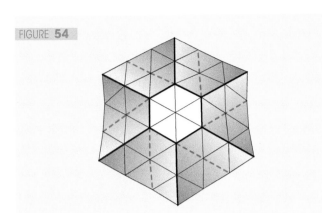

FIGURE **54**

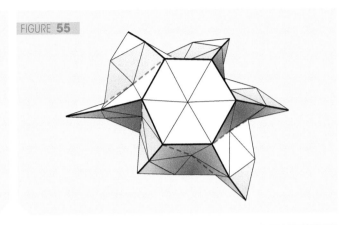

FIGURE **55**

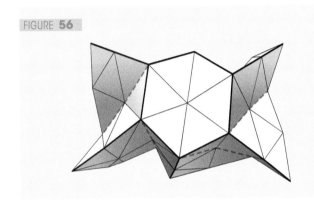

FIGURE **56**

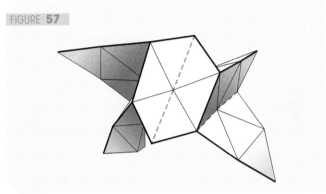

FIGURE **57**

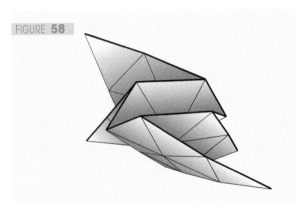

FIGURE **58**

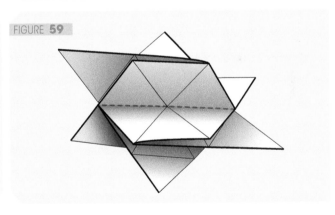

FIGURE **59**

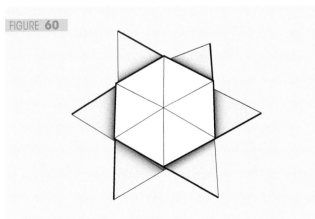

FIGURE **60**

# Rhombus Twist

The rhombus twist is a somewhat quirky fold: This twist is made up of two triangle twists that are so close to each other that they connect—and make a rhombus instead!

**1** Start by folding staggered, equally spaced creases on a precreased triangle grid, as shown in figure 61. Pay close attention to the additional mountain-fold creases you will need to create to connect the four points of the rhombus.

**2** Collapse the twist along the valley folds (figure 62), which will help the rhombus start twisting into shape (figure 63).

**3** Finish the rhombus twist by making sure each pleat is folded directly above the neighboring pleat (figure 64) and that all the creases line up exactly. This twist can be a tricky one to fold accurately.

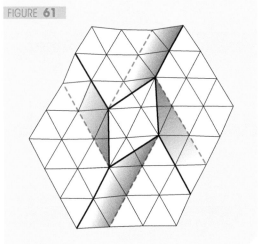

FIGURE **61**

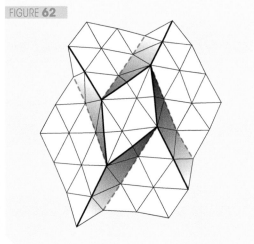

FIGURE **62**

FIGURE **63**

FIGURE **64**

## Open-Back Triangle Twist

The open-back triangle twist is a slight variation from the regular triangle twist. Basically, you shift the pleats over by one pleat width, so instead of meeting together at a single point they overlap a bit. This overlap creates an open triangular space. You create new creases that aren't aligned with the grid, and these new creases form the open triangle in the back.

**1** Fold the crease lines as shown in figure 65, slightly and consistently off-center on a precreased triangle grid. Pay close attention to the creases you are creating for your newly formed triangle, so the folds accurately connect the corners of your triangle together. These folds sometimes are tricky due to the difficult angles involved.

**2** Once the new creases are in place, collapse the twist along the valley folds. Your triangle will be twisted into shape (figure 66).

**3** Finish the open-back triangle twist by making sure the pleats align cleanly with one another (figure 67). The back side of the piece should have a single triangle-shaped hole between the pleats.

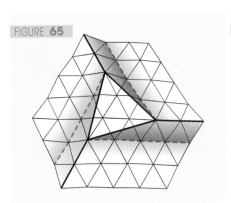

FIGURE **65**

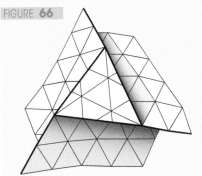

FIGURE **66**

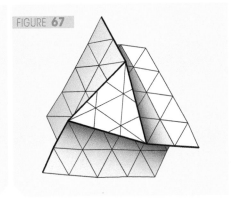

FIGURE **67**

## Open-Back Square Twist

Although it is similar to the square twist, the open-back square twist employs diagonal creases that don't match up with the common 90- and 45-degree angles and lines commonly used in square-based tessellations. Because of these different angles, an open space is created on the back side of the paper opposite the twist.

Open-back square twists differ from regular twists in their inability to change rotational direction without requiring new pleats to be made. You can use different diagonal distances to change the amount of rotation and the overall size of the offset twist.

**1** Fold slightly offset creases on a precreased square grid, as shown in figure 68.

**2** Using these crease lines, start to twist the square shape and rotate it into position (figure 69).

**3** When you fold the pleats over, the square often will follow along and simply snap into place (figure 70). Make sure the new square's pleats align cleanly and consistently.

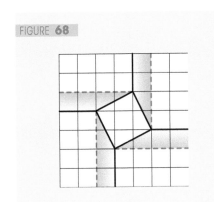

FIGURE **68**

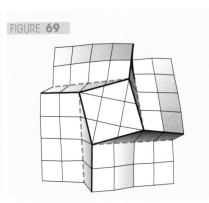

FIGURE **69**

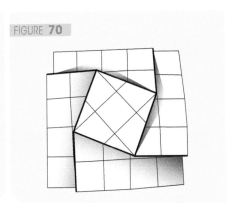

FIGURE **70**

# Open-Back Hexagon Twist

The open-back hexagon, or hex, twist is the most common of all the open-back twists. This twist is easy to fold and facilitates the structure of many designs.

The extreme simplicity of the open-back hex twist relative to other open-back twists may seem counterintuitive, given that a hexagon has six sides—more than a square or a triangle. But the geometry of the hexagon lends itself to many types of manipulations, which accounts for its popularity above all other shapes in the tessellation-folding community.

**1** To fold an open-back hex twist, fold offset creases connected to each point of the six points of a hexagon on a precreased triangle grid, as shown in figure 71.

**2** Once all the creases are in place, collapse the piece along the valley folds, folding the pleats directly over and onto their neighboring pleats (figure 72).

**3** Snap the finished twist into place (figure 73). The back side of the piece will have an open space the size of a one-pleat-wide hexagon.

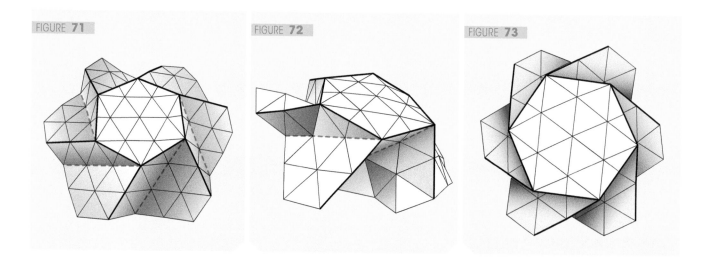

FIGURE **71**   FIGURE **72**   FIGURE **73**

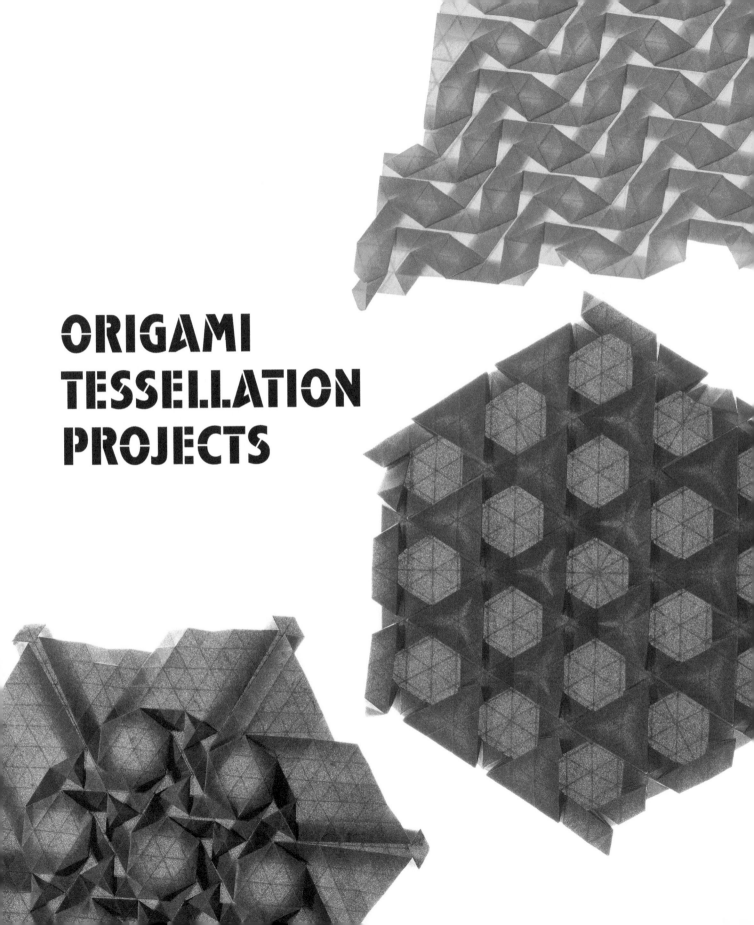

# ORIGAMI
# TESSELLATION
# PROJECTS

# FIVE-AND-FOUR

**This Modernist tiling of squares employs simple 90-degree pleat intersections.
Even the most basic tessellation technique can create complex, attractive designs!**

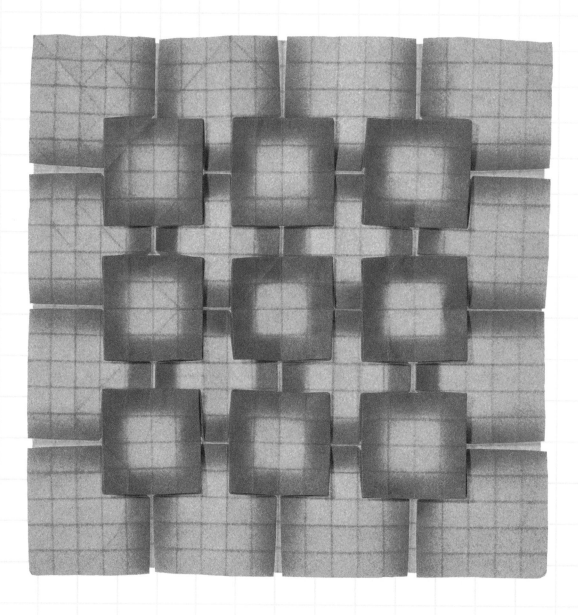

*techniques*
Square Grid, page 6
90-Degree Pleat Intersection, page 11

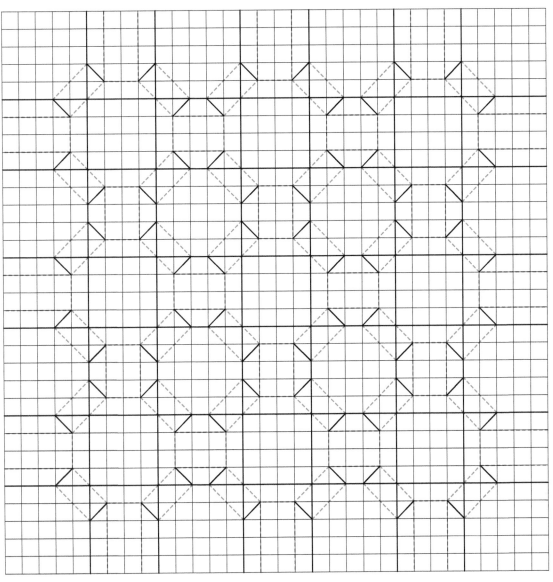

## Instructions

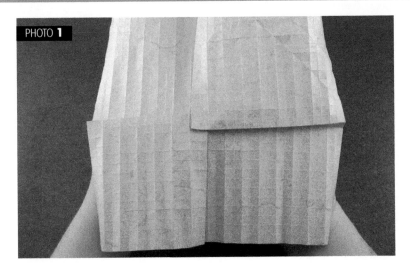

PHOTO **1**

**1** Start with a square grid folded into 32nds. Identify the four-by-four square that's centered in the middle of the paper. This square will become the central tile of the finished design.

**2** Fold two adjacent edges of the square into pleats, making a 90-degree pleat intersection. Keep the radiating pleats folded back, so the paper appears to hold two protruding squares with overlapping corners (see photo 1).

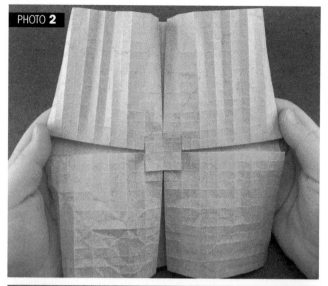

PHOTO **2**

**3** Repeat step 2 to create the next corner of the central square, and then complete the square by pleating the last two corners. The resulting square tile has two facing pleats extending out from the center of each side (see photo 2).

**4** Now comes the tricky part: Unfold one side of the central square tile slightly to open up the two parallel pleats emanating from it. From the edge of the central square, count outward five grid squares and identify another four-by-four square—just like the central square— starting on that row. Make a 90-degree pleat intersection for each corner of the new square, starting with the two corners that directly face the central square. You must fold these two corners first for the paper to lie flat (see photo 3). Then, fold the remaining two corners of the second square.

**5** Rotate the paper 180 degrees and repeat this process to add another square. The paper should now have three square tiles in a central row.

**6** Turn the paper so the three square tiles line up horizontally across the paper. Just as you created a new square next to an existing square in steps 4 and 5, do the same for each of these three squares. Count down five grid squares from the edge of each tile and identify a new square. Fold the two facing pleat intersections first for each of the new squares (see photo 4). Then, fold the outer two pleat intersections to complete each new square tile (see photo 5).

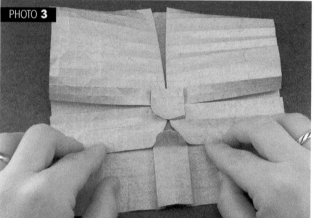

PHOTO **3**

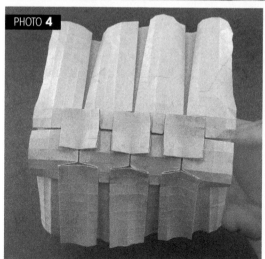

PHOTO **4**

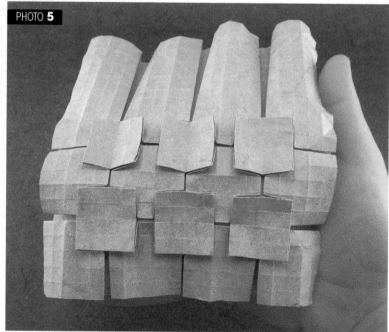

PHOTO **5**

**7** Rotate your paper 180 degrees and create the last three squares in the same way.

Depending on the type of paper you use, your finished piece might curl a bit. To fix the problem, place your folded tessellation between the pages of a large, heavy book. Be sure all the creases and pleats are lying exactly how you want them, because the pressure of the closed book will make the creases permanent.

Leave your tessellation in the book overnight. By morning, you will have a nicely pressed origami tessellation!

## TIP

Accuracy is the key to folding origami tessellations. A lack of precision can derail any project. Remember: Good work going in means good work coming out!

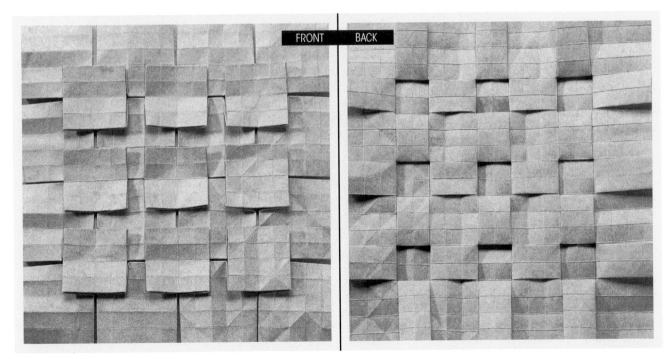

FRONT     BACK

# SPREAD HEXAGONS

In this easy tiling based on a single repeated hexagon, the hexagons circle around the center of the pattern, appearing to be stacked on top of one another. This design was the first origami tessellation I discovered on my own.

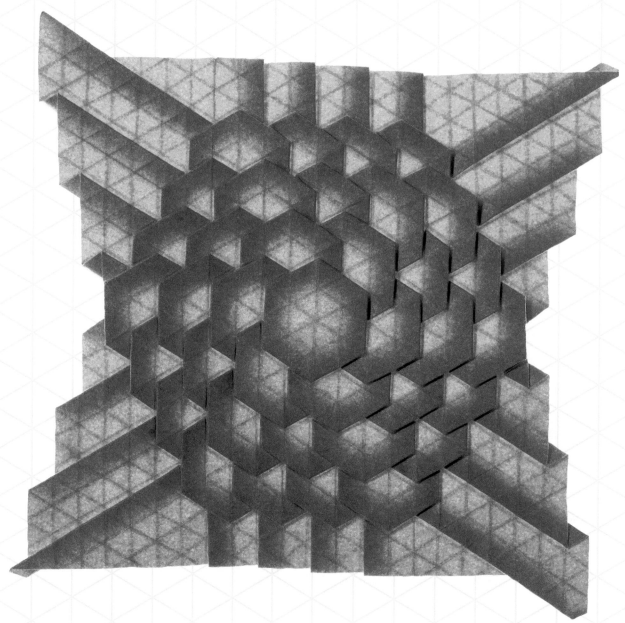

*techniques*
Triangle Grid, page 7
120-Degree Pleat Intersection, page 9

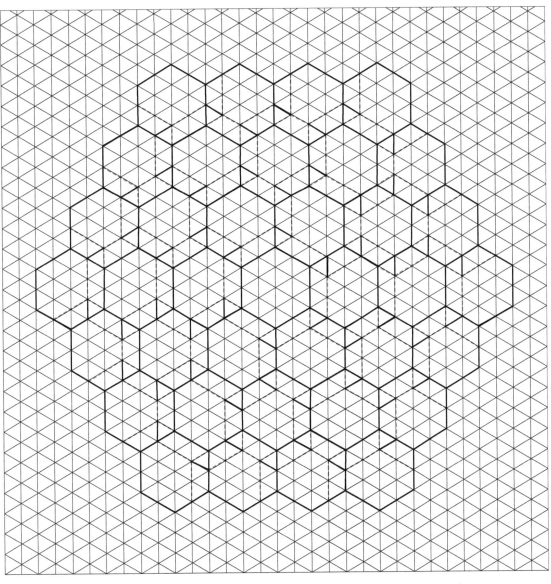

## Instructions

PHOTO **1**

**1** Begin with a precreased triangle grid folded into 16 or 32 divisions. Sixteen divisions will help you understand the idea—and it's a good place to start—but 32 divisions will show off the design to its fullest. Locate the center of the grid, and mark out a hexagon that's two triangle lengths to a side. The resulting hexagon will be four pleats across in all directions.

**2** Fold each corner of the hexagon as a 120-degree pleat intersection. The resulting hexagonal tile in the center of the paper will have pleats radiating out from each corner (see photo 1).

**3** Choose a pleat and count outward two pleat lengths from the hexagon tile. Start counting at the base of the hexagon. Because one pleat length is underneath the hexagonal tile, it will look like you moved only one pleat away (see photo 2). Fold another 120-degree pleat intersection here, making sure to fold the new pleats away from the center. This new pleat intersection is actually the edge of another hexagon tile.

**4** Repeat step 3 for the remaining five sides of the central hexagon. As the new pleats intersect each other farther out on the paper, just fold them over one another—this arrangement is only temporary. When you have completed all six 120-degree pleat intersections, you will see a star-shaped design with the hexagonal tile in the center (see photo 3).

**5** At one point of the star, unfold the pleats a bit. Make two new 120-degree pleat intersections where the tip of the star is two grid triangles wide. This action creates a new hexagon tile, identical in size to the central hexagon and seemingly positioned underneath it. Work around the star, creating five more hexagon tiles (see photo 4).

PHOTO **2**

PHOTO **3**

PHOTO **4**

**6** To expand the pattern, treat each new hexagon tile like the original center tile. Count two pleats outward, fold a 120-degree pleat intersection, and repeat. Fold the overlapping pleat intersections to make a new layer of hexagonal tiles (see photo 5).

Each layer has more and more hexagons as you work your way outward. By maintaining the same pleat orientation, the folds will all flow into each other. You can continue this tessellation into infinity by following the basic folding pattern. See an extended version of *Spread Hexagons* on page 119 in the Gallery section.

PHOTO **5**

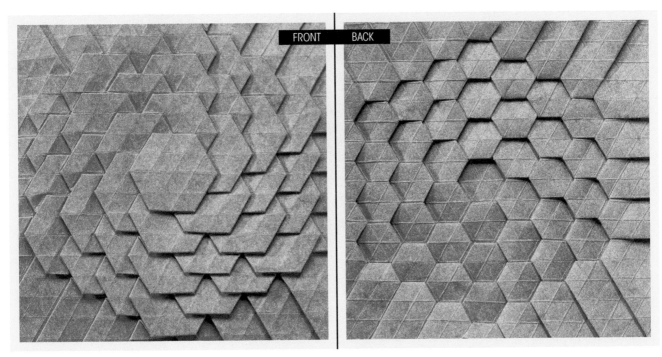

FRONT    BACK

# TILED HEXAGONS

In Tiled Hexagons, the pleats of a basic hexagon tiling are connected by triangle twists—important folds for origamists to learn.

*techniques*
Triangle Grid, page 7
Triangle Twist, page 16

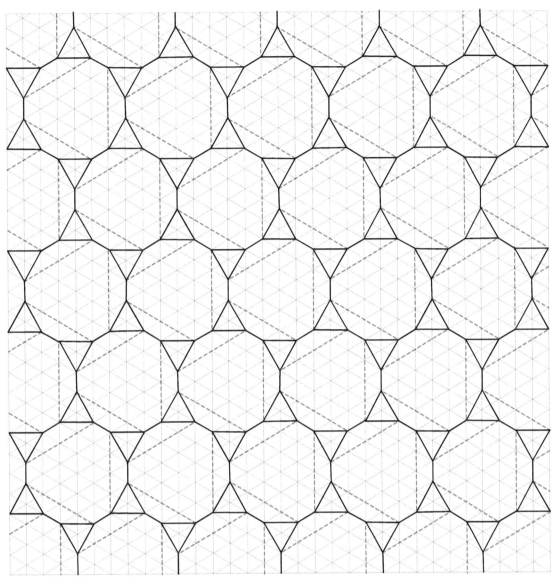

*Instructions*

PHOTO **1**

**1** On a triangle grid, mark out a hexagon three pleats wide on each side centered on the paper.

**2** Create a corner of the hexagon by folding over the pleat extending outward from it, forming a 120-degree pleat intersection. Fold the next corner in the same manner, except in the opposite direction: If you folded the first pleat in a clockwise direction, fold the second counterclockwise. Continue around the hexagon following this pattern (see photo 1).

**3** Fold one side of the hexagon toward the center. This move raises up the 120-degree pleat intersection. Now, squash it flat into a triangle twist (see photo 2). Repeat the squashing all the way around the central hexagon (see photo 3).

**4** Tracking along one of the pleats extending outward, fold a new triangle twist centered two pleat lengths out from the edge of the first twist. (In this design, every pleat intersection should twist, and every twist should be a triangle.) Work your way around to create a new hexagon with triangle twists. The new hexagon will share two triangles with the center hexagon (see photo 4).

**5** Continue outward from the central hexagon in all directions, making more hexagons connected to the center (see photo 5) and then radiating out in orderly fashion from the center. The pattern can continue on to the edge of your paper.

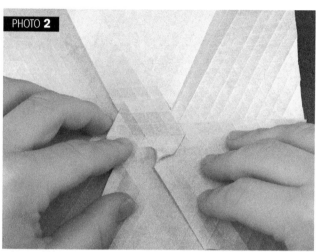

PHOTO **2**

PHOTO **3**

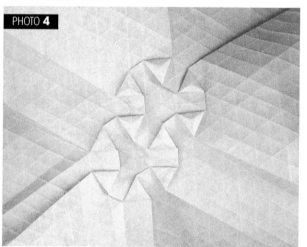

PHOTO **4**

## TIP

The most important folds are your initial reference creases. These folds form the base for the rest of your work. The entire pattern flows from them.

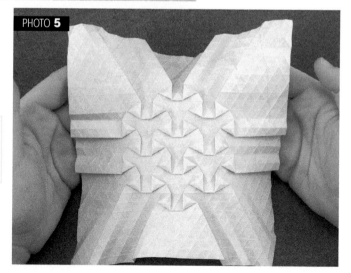

PHOTO **5**

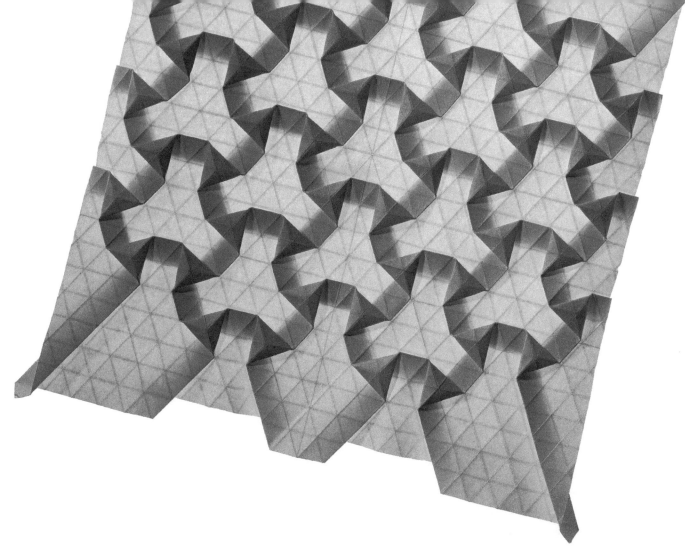

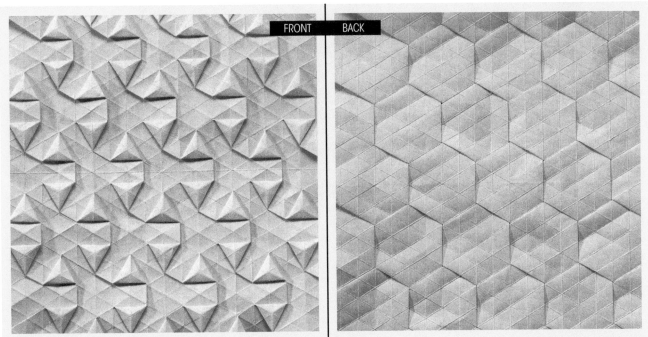

# No.4 STAR PUFF

Designed by Ralf Konrad of Germany, this tessellation enchants with its three-dimensional star shapes. The design is easy to fold and provides a great introduction to the often-curious properties of triangle twists.

*techniques*
Triangle Grid, page 7
Triangle Twist, page 16

## Instructions

PHOTO 1

**1** Fold the Spread Hexagons pattern (see page 28) on a precreased triangle grid with at least 32 divisions.

**2** On the central hexagon, fold a triangle twist at every corner (see photo 1).

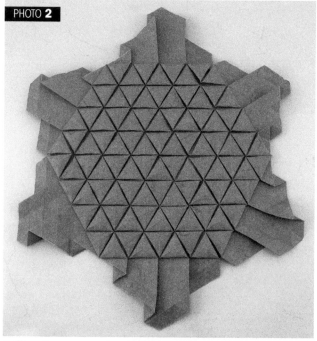

PHOTO 2

**3** Every corner of every hexagon is a 120-degree pleat that can be squashed into a triangle twist. Continue outward from the central hexagon, folding triangle twists at every corner of each adjoining hexagon (see photo 2). Proceed in orderly fashion, folding every corner of a given hexagon before moving to the next hexagon.

**4** Locate the central hexagon, which now is a grouping of triangle twists, to fold your first star. The six twists are composed of three flaps, each with two connected triangle twists. Open these flaps (see photo 3) to create a triangle with three pleats on every side.

**5** Push up on the triangle from the opposite side of the paper to pop up a bubble hexagon. Continue to push up from below with one hand while you push in the sides of the hexagon with your other hand to create star points (see photo 4). The sides should simply pop into place as you create a three-dimensional six-sided star.

**6** Repeat the hexagon opening and popping wherever you want stars in your design (see photo 5). You can create many variations by popping different arrangements within the pattern.

PHOTO 3

PHOTO 4

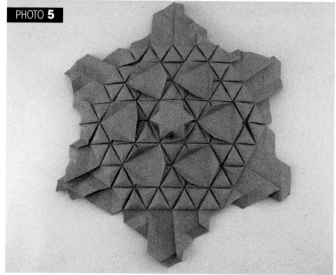

PHOTO 5

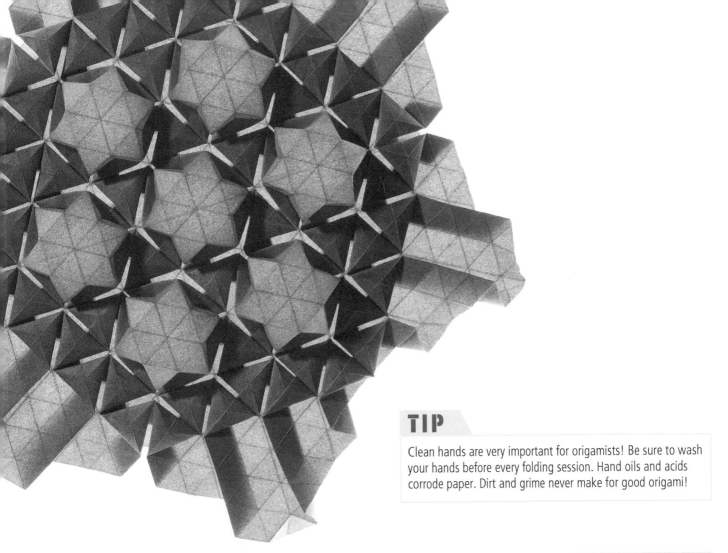

## TIP

Clean hands are very important for origamists! Be sure to wash your hands before every folding session. Hand oils and acids corrode paper. Dirt and grime never make for good origami!

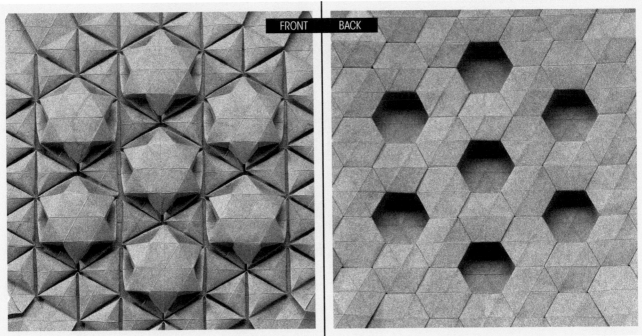

FRONT BACK

# 3.6.3.6

The 3.6.3.6 tessellation is a classic—one of the most popular and frequently folded origami-tessellation designs. First discovered by renowned origamist Shuzo Fujimoto, 3.6.3.6 offers a primer in the rotational direction of twist folds. These twists fold properly in only one direction!

## techniques
Triangle Grid, page 7
Triangle Twist, page 16
Hexagon Twist, page 18

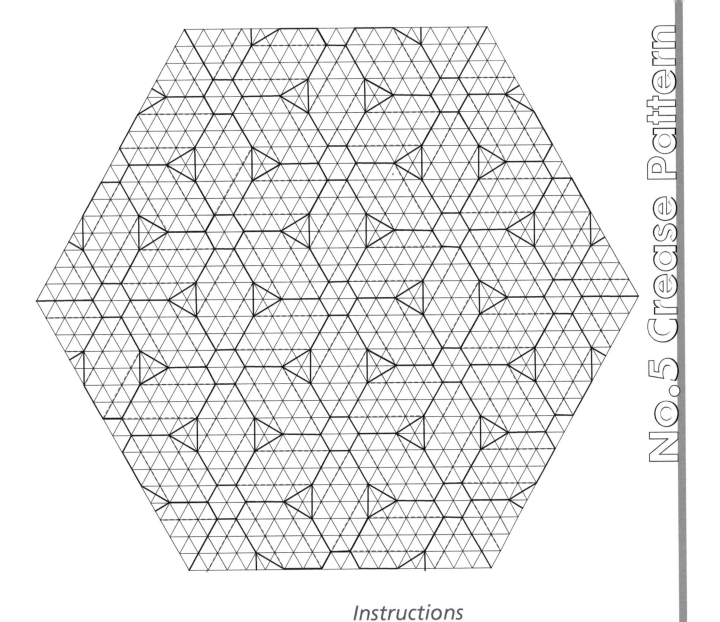

## Instructions

PHOTO 1

**1** On a precreased triangle grid, mark the center point of the page and fold a hexagon twist around it.

**2** Move three pleats outward from the hexagon twist and fold a 120-degree pleat intersection (see photo 1). Remember that one pleat is hidden under the hexagon, so you will see only two pleats between the edge of the hexagon and the new pleat intersection. Continue around the pleats extending from the hexagon twist, folding five more 120-degree intersections.

PHOTO 2

**3** At one 120-degree pleat intersection, fold the pleat up, squashing the resulting shape into a triangle twist (see photo 2). You'll notice an odd bunching of pleats farther out now that you have folded a triangle twist. This pleat intersection is actually another hexagon twist. It just needs to be roughed out and brought into shape (see photo 3).

**4** Move around the central hexagon twist to the next 120-degree intersection and repeat step 3 to make a triangle twist and then a new hexagon twist after it. You'll notice more pleats that do not lie flat and that connect the hexagon twists together. They are triangle twists waiting to be squashed flat (see photo 4). Oblige them!

**5** Work around each successive hexagon twist, folding all the triangle twists until you have folded the entire sheet of paper. As with all tessellations, this pattern tiles infinitely, so once you get the hang of how the basic pattern of hexagon twists and triangle twists comes together, you can extend it out as far as you'd like.

PHOTO 3

PHOTO 4

FRONT    BACK

# PINWHEEL

My trial-and-error experiments folding pinwheel shapes yielded this wonderful tessellated pattern. Easily among my favorite designs, I hope it becomes one of yours, as well.

*techniques*
Triangle Grid, page 7
Hexagon Twist, page 18
Open-Back Triangle Twist, page 21

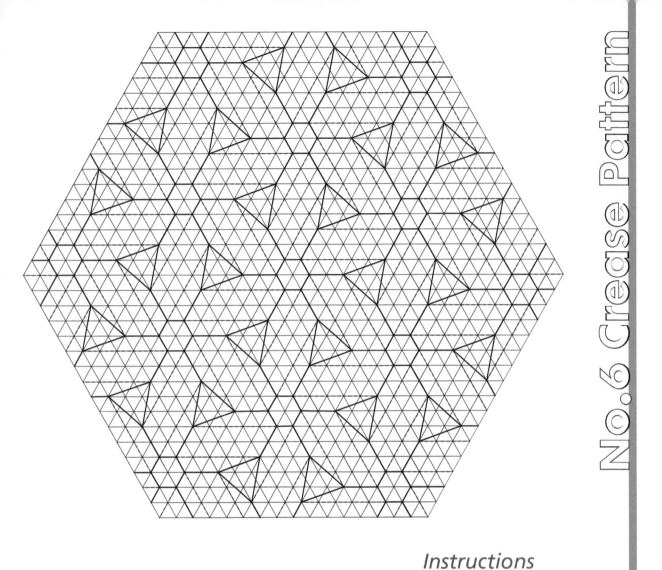

*Instructions*

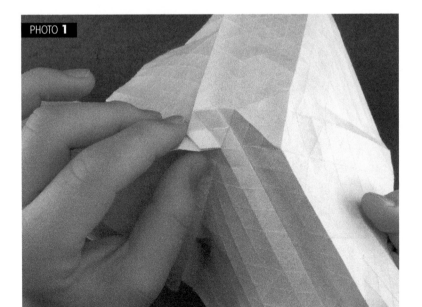

PHOTO **1**

**1** Make a hexagon twist in the center of a precreased triangle grid with at least 32 divisions. Precrease the outlines for one of the triangles shown in the crease pattern. (See photo 1: The triangle is set to the right of the hexagon.) This triangle will form an open-back triangle twist. Collapse the open-back triangle twist, which will fold flat quite easily along the crease lines.

**2** Repeat step 1 for the next pleat radiating from the central hexagon twist. You'll encounter some troublesome pleats that intersect between the two triangle twists. These pleats eventually will become another hexagon twist, but for now fold them flat using the techniques detailed in the 60-degree pleat intersection instructions on page 12. Rather than folding the pleats toward each other, fold both over in the same direction (see photo 2).

**3** Continue around the central hexagon twist, making four more open-back triangle twists and, to keep things tidy, flattening their related pleat intersections (see photo 3). Although you can leave the pleats loose, you'll be folding them into hexagon twists shortly, so it's best to lay some of the groundwork ahead of time.

**4** Turn your attention to one of the partial hexagon twists formed by the "troublesome pleats" of steps 2 and 3. You've already folded two of the six corners! At one of the unfolded corners, form a 120-degree pleat intersection—refer to the crease pattern for location guidance, if you need help. Repeat the step on an adjacent hexagon-twist-to-be, and form an open-back triangle twist between the two (see photo 4). Again, use the crease pattern as your guide.

**5** Work your way around the hexagon twist being formed, creating pleats and folding open-back triangle twists. When you're finished, make sure the twists are fully flat. You might occasionally need to unfold one a bit and twist it in the opposite direction for it to fold completely. A good fold makes a nice, clean twist closure on the reverse side of the paper.

**6** Repeating steps 4 and 5, work through the remaining areas of the paper, folding open-back triangle twists and hexagon twists with abandon (see photo 5)!

PHOTO **2**

PHOTO **4**

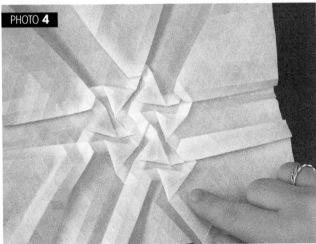

PHOTO **3**

PHOTO **5**

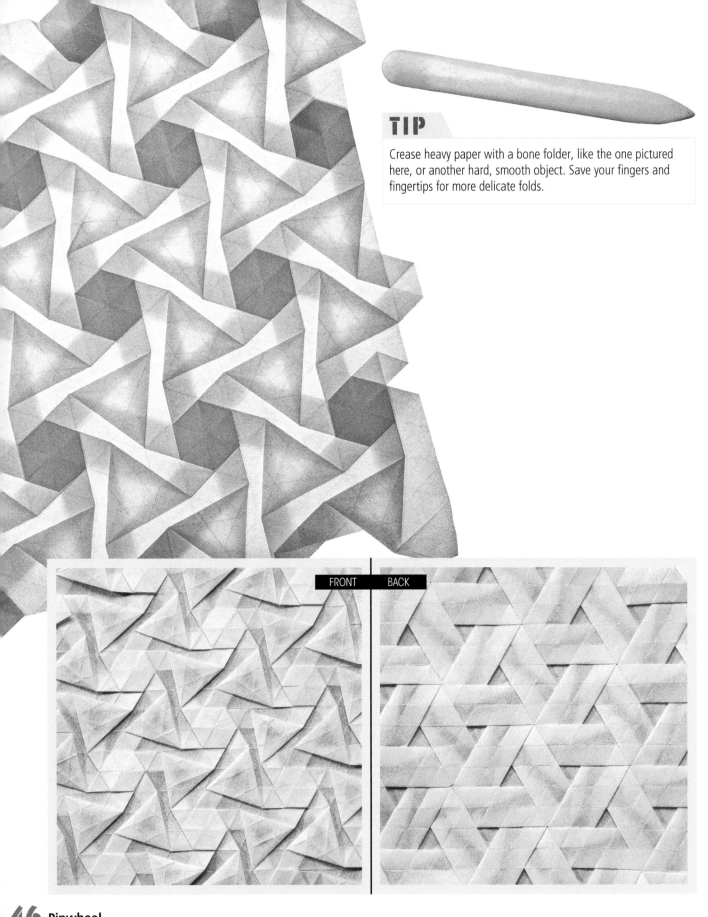

FRONT    BACK

# OPEN-BACK HEXAGON TWIST

**No. 7**

Closely related to the 3.6.3.6 tessellation, this design creates open-back hexagon twists, which give it a very different look and feel. Even though the underlying patterns are similar between the two tessellations, a slight modification to Open-Back Hexagon Twist dramatically alters the final piece. Once you complete the design, your biggest question will be which spectacular side to display.

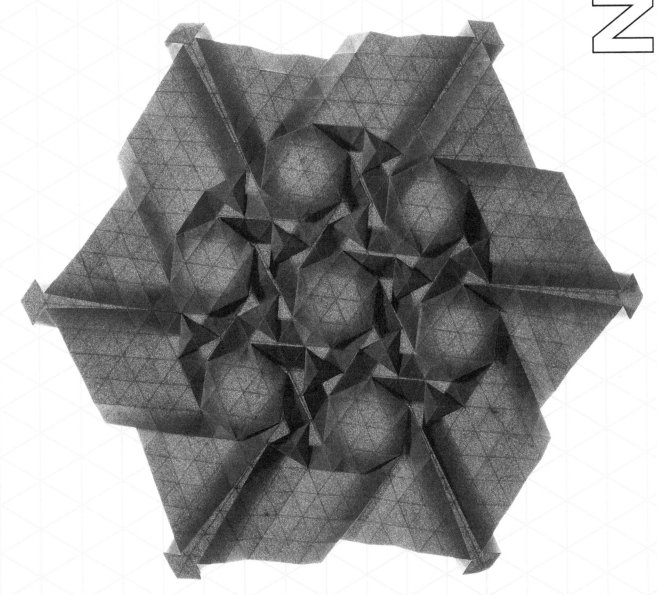

*techniques*

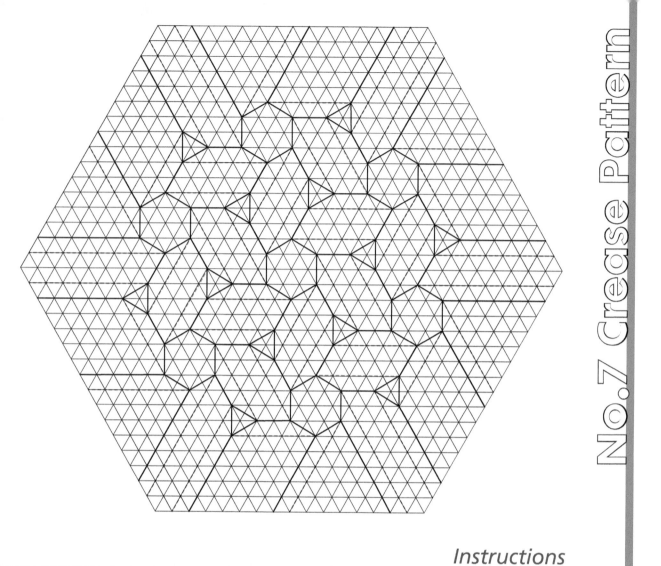

## Instructions

PHOTO **1**

**1** Fold an open-back hexagon twist in the center of a triangle grid. Count outward three pleats along the pleat line extending from one corner of the open-back hex twist, and fold a triangle twist centered on the third pleat (see photo 1).

**2** Move to the next hexagon corner and repeat step 1. You'll notice the pleats extending from the triangle twists interfere with one another. Where they intersect, fold another open-back hexagon twist. This new open-back hex twist will connect the two pleats, pinching into place an offset line from a point two pleats away from the tip of each triangle twist (see photo 2).

**3** Return to the central open-back hex twist. Move to the next radiating pleat, and make another triangle twist. Create another open-back hex twist, and fold an additional triangle twist between the two secondary open-back hex twists where their radiating pleats intersect (see photo 3).

**4** Continue working around the central hexagon, folding triangle twists and open-back hexagon twists. Your final design will be a field of hexagon twists surrounded by triangle twists, and the reverse side of the piece will feature a beautiful, interlaced grid of set-back hexagons.

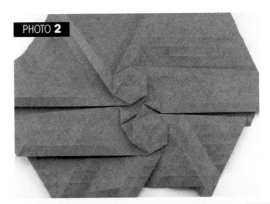

PHOTO **2**

PHOTO **3**

FRONT · BACK

# BASKET WEAVE

Building on the 3.6.3.6 and Open-Back Hexagon Twist tessellations, Basket Weave packs its twists so closely together they actually overlap one another. The tessellation takes on a natural curve and dimensionality during folding. The final piece resembles a woven basket made of rattan. A common design, I was first introduced to this pattern by Joel Cooper.

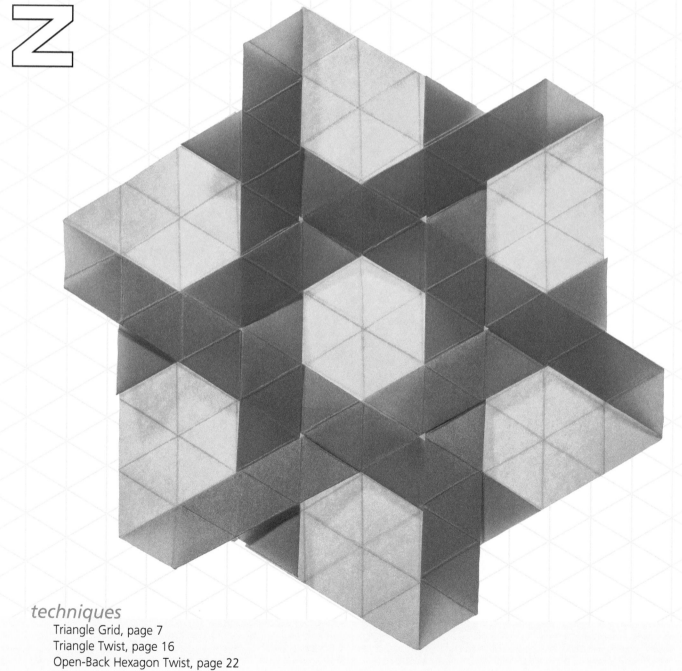

*techniques*
Triangle Grid, page 7
Triangle Twist, page 16
Open-Back Hexagon Twist, page 22

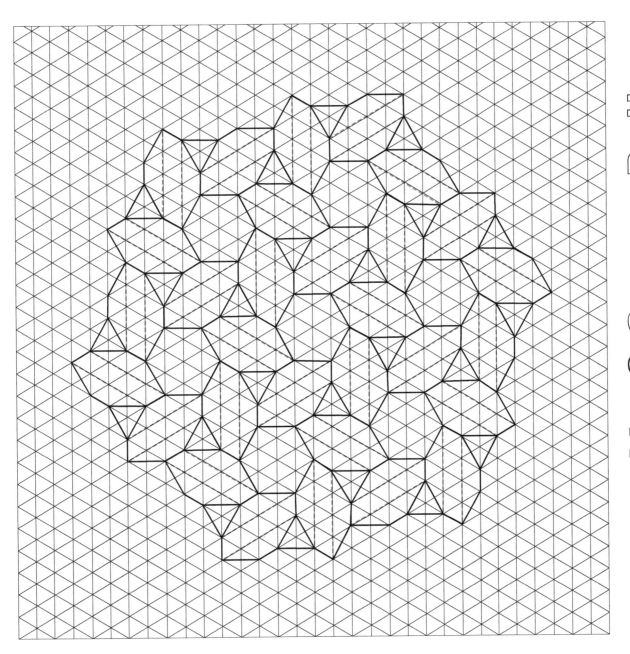

## TIP

When choosing your paper, make sure the piece is large enough for the pattern you have in mind. Think ahead! Consider how wide your final pleats will be. A little planning will prevent a lot of frustration.

# Instructions

**1** Start with a triangle grid, and fold an open-back hexagon twist in the center.

**2** Moving outward one pleat from the edge of the hexagon twist, fold a regular triangle twist (see photos 1 and 2). The triangle twist will overlap the initial hexagon twist. Repeat this process for all sides of the hexagon.

**3** Now the folding starts to get a little tricky! Fold open-backed hexagon twists where the pleats from the triangle twists come together. See photo 3, which isolates folds on one side of the hexagon, and photo 4, which illustrates how to "tuck under" these folds.

**4** The flip side of the tessellation—which I consider the front side—is starting to develop a wonderful basket-weave pattern (see photo 5). Look at this side to verify the accuracy of your work. You should not see any double-wide pleats or other shapes—only hexagons and thin strips of single-width pleats that all line up cleanly.

PHOTO **1**

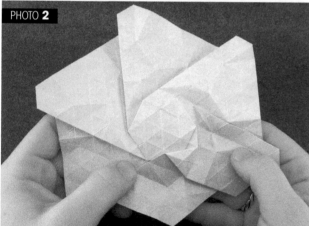

PHOTO **2**

PHOTO **3**

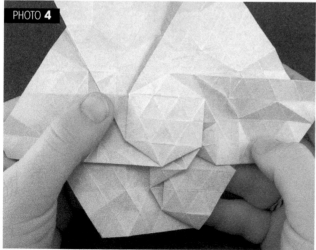

PHOTO **4**

PHOTO **5**

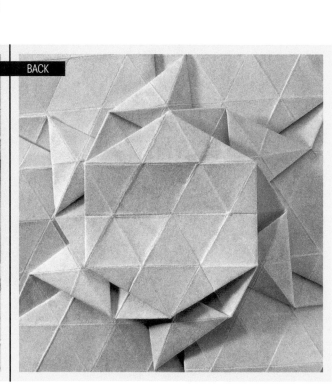

**5** Once completed, the back side of the tessellation will have many layers of twists piled on top of one another (see photo 6). Try experimenting to discover which layering scheme you like best.

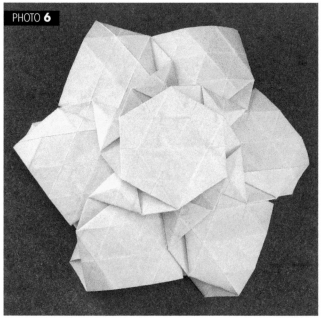

PHOTO **6**

FRONT   BACK

# WATER BOMB

The Water Bomb tessellation is made from one of the most popular bases in origami. By tiling a simple base with squares, you achieve a three-dimensional surface that's both pliant and curving. The larger you scale the pattern, the more the surface of the paper curves.

*techniques*
Square Grid, page 6

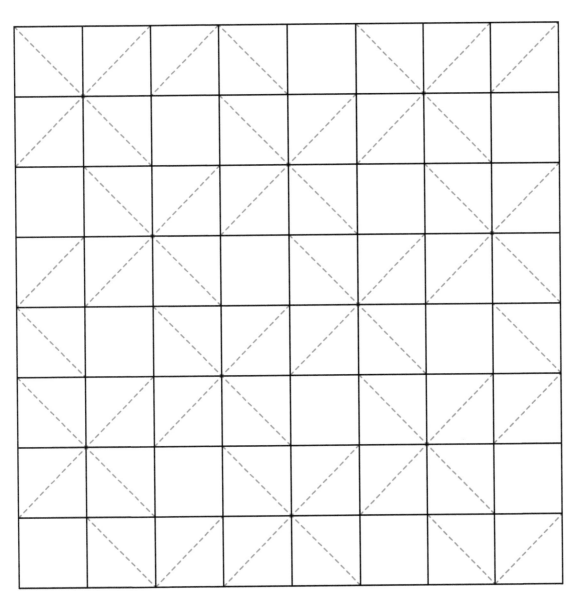

## Instructions

PHOTO 1

**1** For best results, use a relatively stiff paper. On an eight-by-eight square grid, precrease all the valley folds as shown on the crease pattern. The easiest way to do this is to flip the paper over and then create mountain folds.

**2** Remaining on the paper's "back side," start at one of the corners and collapse the folds together along the new creases (see photo 1).

**3** Loosely pinch in all the folds, so they take on a slight three-dimensional shape.

**4** Go back to the corners and start collapsing the Water Bomb bases together more tightly.

**Water Bomb** 55

PHOTO **2**

This action requires you to hold the folded portions together while folding the rest (see photo 2). After a certain point, the remaining folds will pop into place almost on their own.

**5** Lay the tessellation down on a flat surface. The tension of the paper forces the pattern apart, which causes it to curve, particularly toward the corners. Hold the paper down at the corners and push the pattern together. Pinch the vertical folds tightly, one by one, to reinforce the precreased folds and to give the paper some "memory" (see photo 3).

**6** You can use paper clips or other suitable clamps to keep the folded pattern together. A Water Bomb tessellation clipped to hold its shape overnight will remember that shape for a very long time!

See an extended version of the *Water Bomb* tessellation on page 118 in the Gallery section.

PHOTO **3**

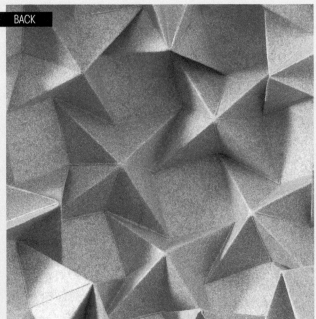
FRONT     BACK

# SQUARE WEAVE

The Square Weave project is a terrific introduction to offset square twists. The finished model is attractive on both sides, featuring a woven pattern on the front and sets of square twists on the back that appear to shimmy across the paper. Please note: Dollar bills folded into square weaves make popular tips at restaurants!

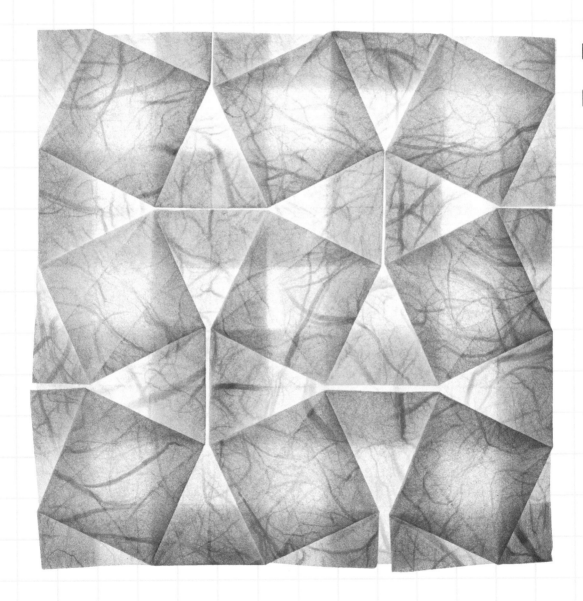

*techniques*
Square Grid, page 6
Open-Back Square Twist, page 21

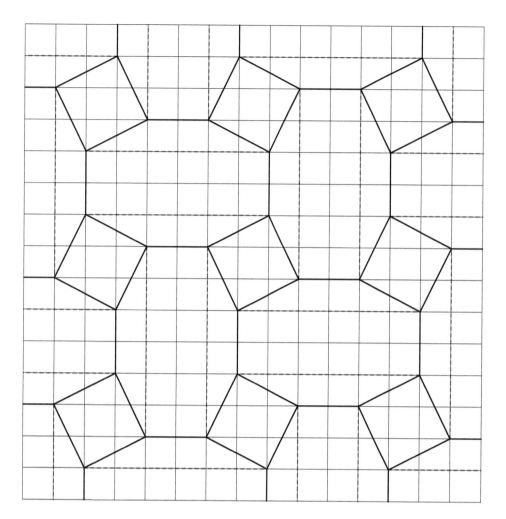

## Instructions

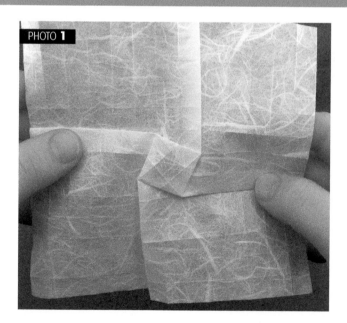

PHOTO **1**

**1** When you first fold this tessellation, I recommend starting with a 15-by-15 square grid—a 16-by-16 grid with two adjacent sides trimmed off. This size grid allows the square weave to perfectly fit three open-back square twists across the paper. (Using a square grid of 32 divisions will give you enough room for six open-back, or offset, square twists across the paper.)

**2** Fold an open-back square twist centered in the middle of the paper (see photo 1). Use the crease pattern for reference if you're not sure exactly where to place the first twist.

**3** Fold the second twist in the opposite direction from the first. If you folded the first one with a clockwise tilt (as in photo 1), fold the second twist counterclockwise. Follow the pleat running to the right from your central twist, and count outward two grid squares. Locate the second square twist so you have two pleat spaces between the respective squares' corners (see photo 2).

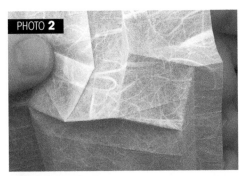
PHOTO 2

**4** Once the second twist is in place, move two pleat spaces off the corner opposite the corner on which you started your second square. Repeat step 3, folding another square twist. You now have three offset square twists, lined up across the center of the paper (see photo 3).

**5** Every offset square twist in this design rotates in the opposite direction of its immediate neighbors. Utilize this understanding in placing the next three twists on the top third of the paper (see photo 4).

**6** Finish the design by rotating the paper 180 degrees and repeating step 5.

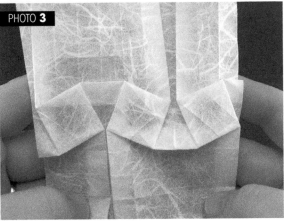
PHOTO 3

PHOTO 4

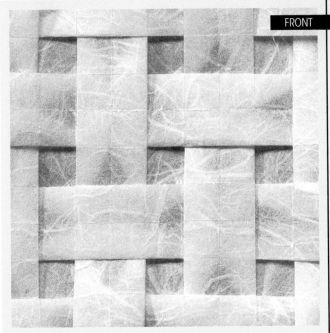

FRONT     BACK

# CHÂTEAU-CHINON

**Inspired by an architectural tiling discovered in the French city Château-Chinon, this distinguished octagon-based design by Christiane Bettens provides delight on both sides of the finished tessellation. Choose a translucent or light paper, because a beautiful pattern appears when the piece is backlit.**

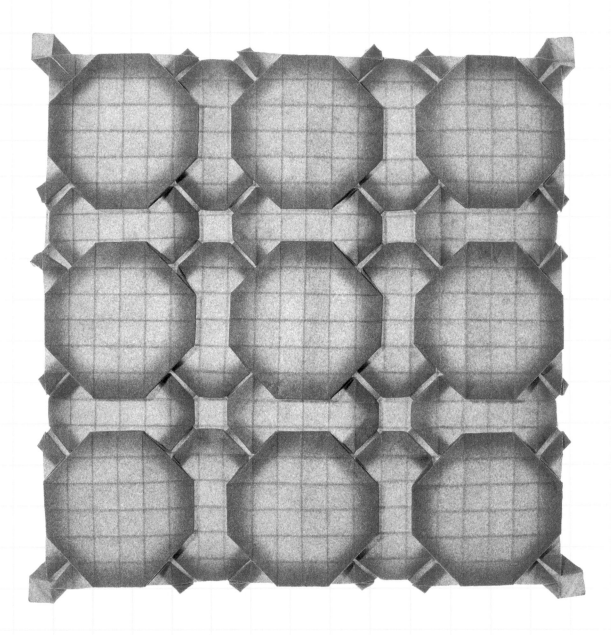

*techniques*
Square Grid, page 6

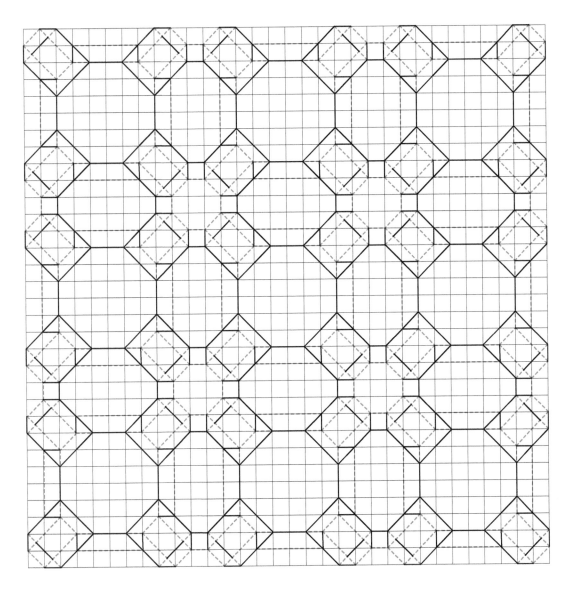

*Instructions*

PHOTO **1**

**1** On a 32-by-32 square grid, fold the diagonal crease lines indicated on the crease pattern. This tessellation has quite a few diagonal creases, and working the piece will be much easier if you fold them ahead of time rather than as you go along. Make sure the valley- and mountain-fold orientations are correct, because the design won't fold together properly otherwise (see photo 1).

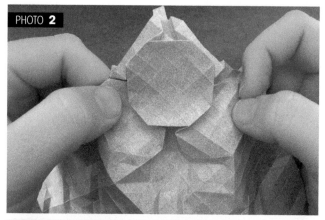

PHOTO 2

PHOTO 3

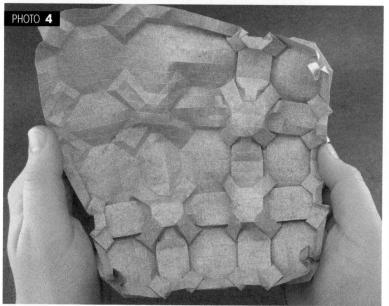

PHOTO 4

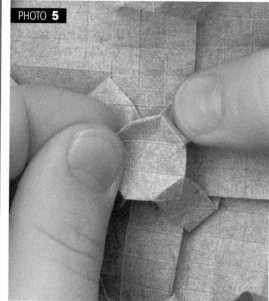

PHOTO 5

**2** Choose one of the piece's corners and start collapsing the design together along the crease lines (see photo 2).

**3** The pattern consists of large octagons, with smaller spaces in between. Don't worry about folding everything completely flat at first. Rough the folds into shape, which will allow more of the design to fall into place as you work your way around the edges (see photo 3). When you're close to finishing, you can fold the paper a bit more tightly—you'll get less resistance from the paper.

**4** Flip the piece over to complete your folding on the back side. The octagon connection points between the main folded sections on the front will be standing out, resembling puffy X shapes (see photo 4). Fold these connection points flat by grabbing the edges and pushing the extra paper inward while pressing the paper flat. The paper should collapse relatively cleanly. Once flattened, each point will look like a small octagon lying on top of an X (see photo 5).

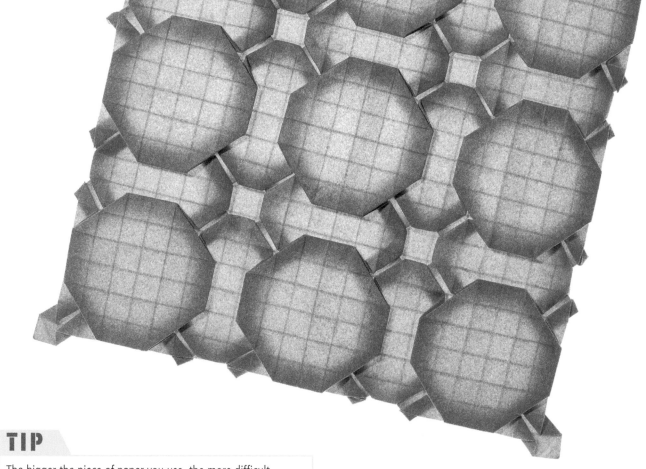

## TIP

The bigger the piece of paper you use, the more difficult folding is. Work your way up to working with larger paper once you have a lot of practice folding smaller pieces.

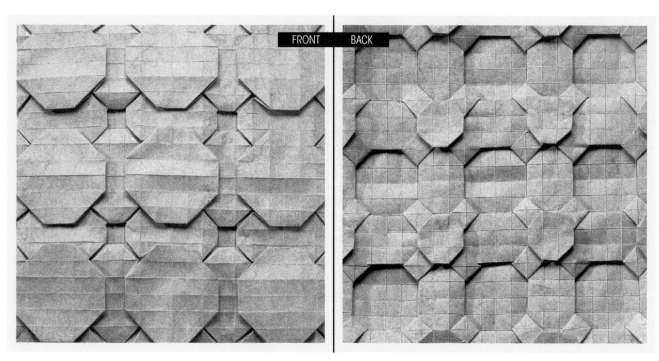

FRONT    BACK

# RHOMBUS WEAVE

Using only one shape—the elegant rhombus—this piece is evocative of herringbone, snakeskin, waves on the sea, and countless other familiar patterns from everyday life.

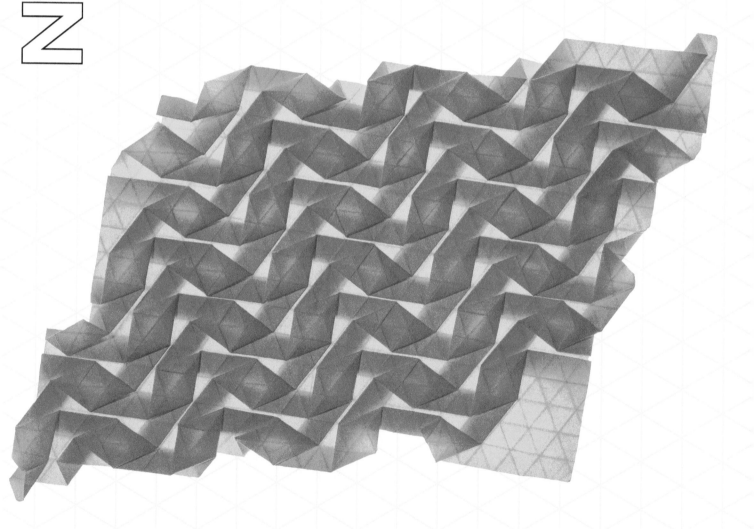

*techniques*
Triangle Grid, page 7
Rhombus Twist, page 20

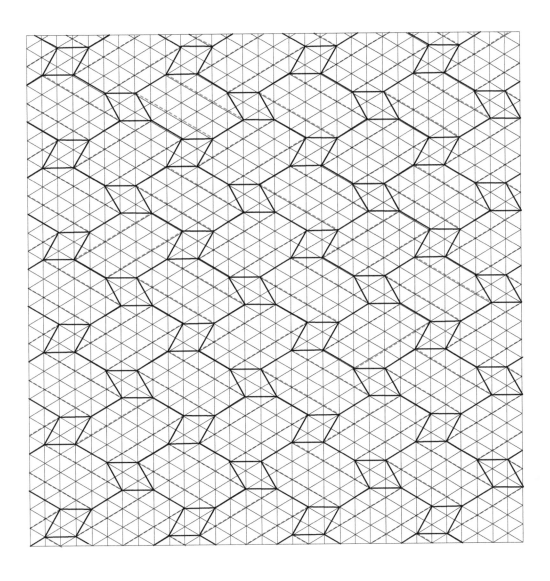

## Instructions

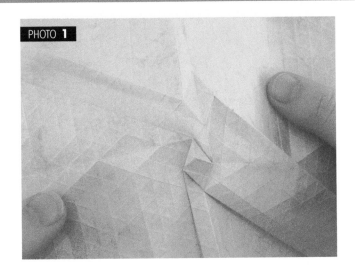

PHOTO 1

**1** Fold a rhombus twist on a triangle grid somewhere near the center of the paper. Unlike many of the projects in this book, where you start isn't critical. Because the entire design is made up of tiled rhombi, the pattern blends together.

**2** Follow a pleat extending out of the first rhombus for two grid triangles, and then pinch in the creases for another rhombus (see photo 1).

**3** Collapse the second rhombus twist so it folds flat. This twist will point in a different direction from the first one. Rhombus twists in this design have only two orientations, and now you have folded both of them.

**PHOTO 2**

**PHOTO 3**

**PHOTO 4**

**4** Moving two grid triangles away from the tip of the previous rhombus twist, make the creases for another rhombus twist (see photo 2). The precreased rhombus points in the same direction as the rhombus you just folded, but when you twist it, the new rhombus reorients in a different direction.

**5** After collapsing the newest twist, you'll notice the pattern essentially consists of a series of steps moving up and down across the paper. Keep folding twists, following this simple pattern, until you have folded the full length of the paper (see photo 3).

**6** Starting a new row is slightly complicated by the existing pleat lines running across the paper. However, these lines also make it much easier to see exactly where your new twists need to go. Think of the pleats as helpful guides rather than pesky intruders! New twists still follow the same rule: Crease lines for new rhombus twists point in the same direction as the previous twist. Start another row, and move along sequentially for best results (see photo 4).

**7** Once you've completed a new row, keep going and fill the remaining paper with rhombus twists. If you started with a square piece of paper, your finished tessellation will be in the shape of a rhombus. Curious, isn't it?

FRONT  BACK

# STAR TWIST

Does this tessellation look like a star or a snowflake? Either way, its simplicity and symmetry trigger natural wonder. The pattern lends itself to easy modification, experimentation, and innovation.

*techniques*

Triangle Grid, page 7
120-Degree Pleat Intersection, page 9
Hexagon Twist, page 18
60-Degree Pleat Intersection, page 12

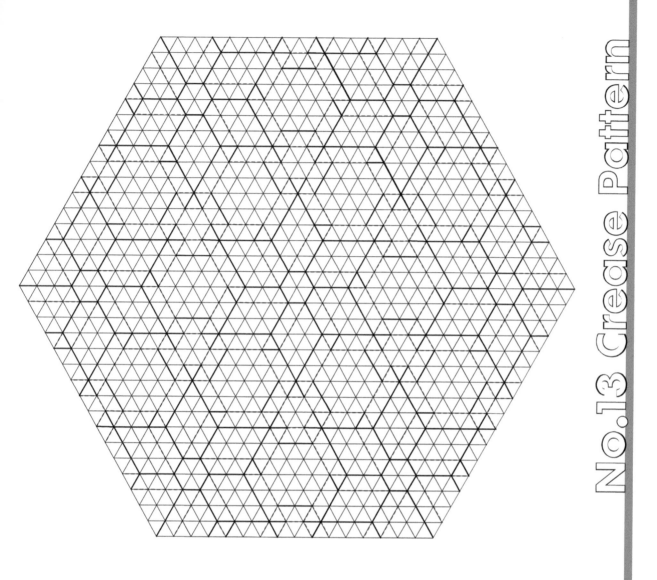

## Instructions

PHOTO 1

**1** Utilizing a triangle grid with divisions of 32 pleats or more, fold a hexagon twist at the center of your paper.

**2** Follow one of the radiating pleats outward for two pleats, and fold a 120-degree pleat intersection. Make sure the two outer pleats are facing away from the central hexagon twist (see photo 1).

**3** Move to the next pleat radiating out from the hexagon twist, and fold another 120-degree pleat intersection two pleat lines out (see photo 2).

**4** Where the pleats from your first intersection and your second one meet, make a 60-degree pleat intersection (see photo 3 showing the back side of the piece).

**5** Work your way around the central hexagon twist, repeating the pleating and intersecting steps until you have completed all six sides (see photo 4).

The easiest way to tile the basic star twist is to treat the 60-degree intersection folds at the tips of the star as one third of a hexagon twist and fold two more star twists, all connecting at that central point. The crease pattern reflects this design.

See an extended version of the *Star Twist* tessellation on page 118 in the Gallery section.

PHOTO **2**

PHOTO **3**

PHOTO **4**

## TIP

Don't force the paper. Gentle encouragement and light folding create better origami than does brute force.

FRONT   BACK

# ROMAN CHURCH FLOOR TILING

**On a trip to Rome, the designer happened upon a gorgeous tiling pattern on an ancient church floor. Inspired by her discovery, Christiane Bettens makes use of numerous triangle twists and double pleats in this pattern.**

*techniques*
Triangle Grid, page 7
120-Degree Pleat Intersection, page 9
Hexagon Twist, page 18
Triangle Twist, page 16

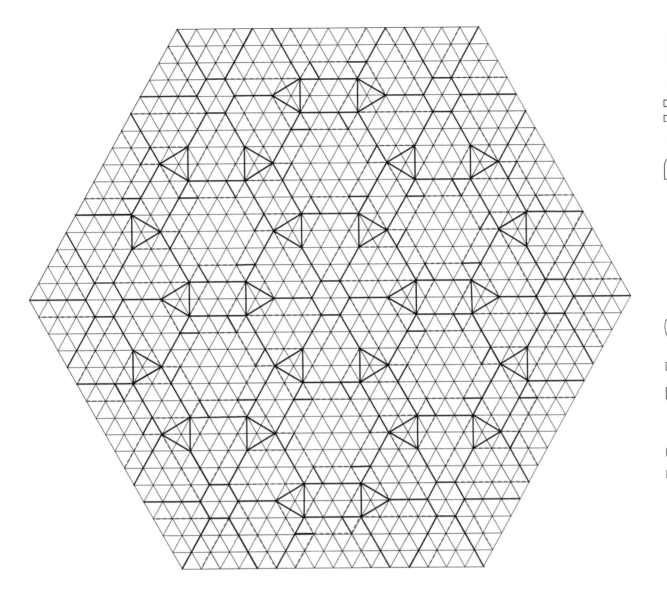

## Instructions

PHOTO **1**

**1** On a large triangle grid, fold a central hexagon twist and then triangle twists around it (see photo 1).

**2** For each triangle twist, fold a 120-degree pleat intersection directly on the two tips of the triangle twist that are not connected to the central hexagon twist. Treat the triangle twist itself as one of the pleats, essentially. Rather than having a single pleat extending out from the corner of the triangle twist, you will have two pleats running 120 degrees apart from each other away from the triangle twist, creating double pleats between the triangle twists (see photo 2). Use the crease pattern to check your folding. Although it may sound complicated, you will find double pleats quite easy to fold.

**3** Moving outward along one of the double pleats extending from the center, fold a triangle twist using the double pleats (see photo 3). As with a regular triangle twist, rotate the paper and squash the triangle twist flat. The triangle twists should touch one another, tip to tip.

**4** Continue around the central hexagon, folding double pleats and triangle twists. This pattern is very similar to the Tiled Hexagons tessellation on page 32, but it uses double pleats instead of single pleats. The double pleats allow you to place hexagon twists in the design because of the extra paper they make available (see photo 4).

**5** Work around the center of the design, folding hexagonal cells made from double pleats and triangle twists. Use the crease pattern for reference to locate the next set of hexagon twists once you have created a complete ring of open hexagonal cells around the center.

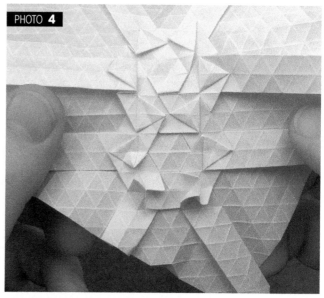
PHOTO **4**

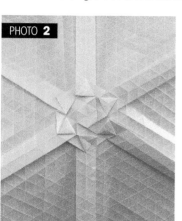
PHOTO **2**

PHOTO **3**

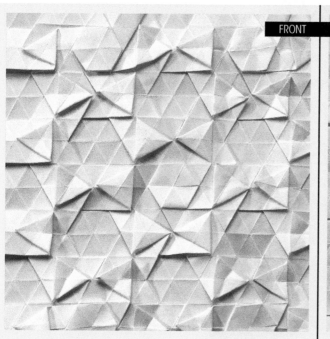

FRONT    BACK

# DAISY CHAINS

A field of flowers folded from a single sheet of paper, this design opens the gate to a wide array of floral modifications. Do you have green thumbs—and eight other green fingers?

## *techniques*

Triangle Grid, page 7
120-Degree Pleat Intersection, page 9
120-Degree Inverted Pleat Intersection, page 10
Hexagon Twist, page 18

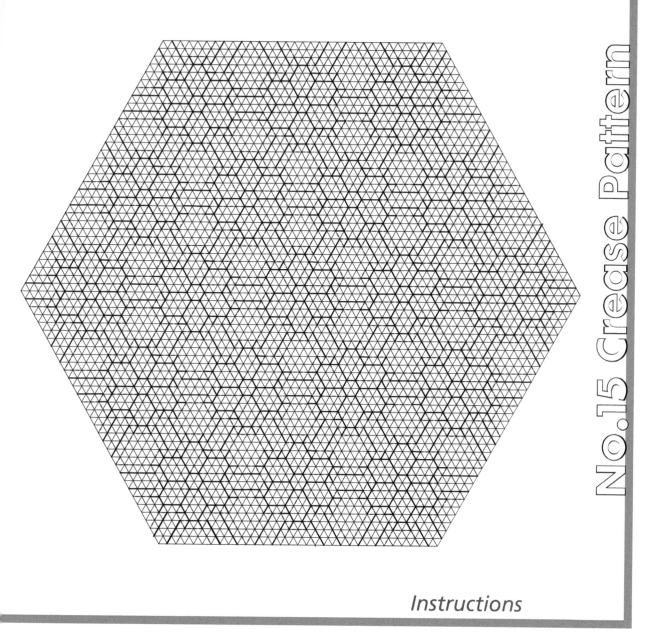

## Instructions

PHOTO 1

**1** Using a triangle grid of 48 divisions or larger, preferably cut into a hexagonal shape, create a hexagon twist in the center of the grid.

**2** Following one of the radiating pleats outward two pleats from the corner of the hexagon, fold a 120-degree pleat intersection (see photo 1).

**3** Fold another 120-degree pleat intersection along the next pleat radiating out from the central hexagon twist. As with other patterns, this radiating pleat will intersect with the first one, and they won't fold flat easily.

**4** Fold down the tip where the two pleats intersect to create a flat-ended petal shape. Essentially, you are folding two 120-degree inverted pleats right next to each other (see photo 2). Refer to the crease pattern for the specific crease locations.

**5** Once you've folded the petal tip and its inverted pleats, the piece should lie flat (see photo 3). Repeat the process for the remaining five sides of the central hexagon. When you've folded all six sides, you have completed one tessellating unit of the design (see photo 4). The pattern can repeat infinitely, of course. The crease pattern reflects the spacing for the design shown in the finished piece. But by changing the spacing between flowers, you can experiment with different results!

PHOTO **2**

PHOTO **4**

PHOTO **3**

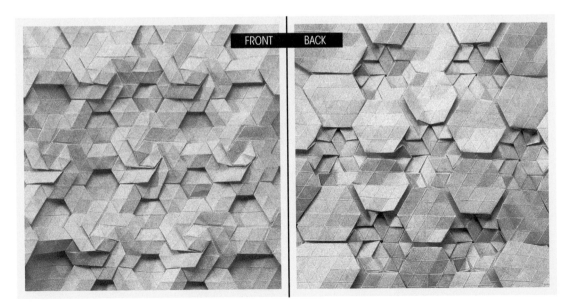

FRONT    BACK

# DOUBLE TRIANGLE SAWTOOTH

**This complex design by Miguel Angel Blanco Muñoz of Barcelona, Spain, features quite a few closely packed twists and pleat intersections. The tessellation is tricky to fold, requires deep patience, and rewards with a dancing, intricate result.**

*techniques*

Triangle Grid, page 7
120-Degree Pleat Intersection, page 9
120-Degree Inverted Pleat Intersection, page 10
Triangle Twist, page 16
Hexagon Twist, page 18

## Instructions

PHOTO **1**

**1** On a triangle grid of at least 32 pleats, identify the center of the paper and fold a hexagon twist.

**2** Employ the 120-degree inverted pleat intersection method to invert each pleat extending from the central hexagon twist. The result is a hexagon twist with pleats extending outward from each corner—often called a "petal fold."

**3** Measuring out along one of the newly created pleats, fold a triangle twist centered one grid triangle from the edge of the central hexagon. The triangle twist should touch the corner of the central hexagon twist (see photo 1).

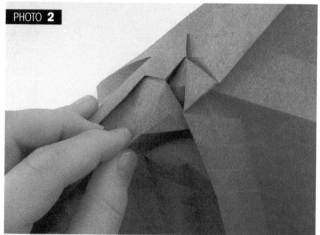

**PHOTO 2**

**4** Repeat the process on the next pleat in a clockwise direction (see photo 2). The second triangle twist will have pleats that intersect with the first twist and so won't lie flat. To solve this problem, unfold the first triangle twist a bit, and fold another triangle twist directly connected to it. The triangles will rotate in opposite directions (see photo 3).

**5** The pleat extending from the adjacent triangle twist (moving clockwise around the central hexagon) will intersect with the lower tip of the bottom-most triangle twist you just folded. Fold a 120-degree pleat intersection to create a pleat leading off toward the corner of the paper. Now the triangle twists all will fold flat (see photo 4).

**PHOTO 3**

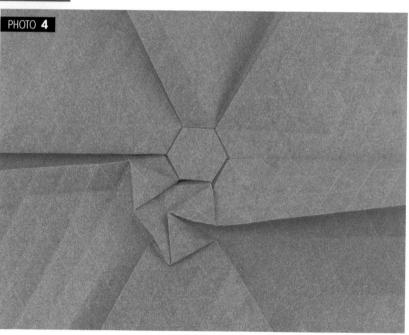

**PHOTO 4**

## TIP

For projects where all the folds of the finished piece lie flat, press your tessellations between two heavy books to prepare them for display.

**6** Move around the central hexagon to the next radiating pleat, and fold another triangle twist next to the hexagon. Repeat the process of creating a second triangle twist to connect with the pleat extending from the new triangle twist (see photo 5). Continue this pattern on all the sides of the central hexagon (see photo 6). The result is the basic double-triangle-sawtooth unit—a wonderful fold in its own right that's even better when tessellated!

**7** To tessellate the pattern, move down along the radiating sets of pleats and fold the double-triangle-twist structures indicated by the crease pattern. As the second sawtooth unit starts to take shape, quite a few pleats will interfere with each other temporarily. Don't try to squash them to make them fold flat. They all will fall into place as you continue to fold the pattern.

PHOTO **5**

PHOTO **6**

FRONT     BACK

# STACKED TRIANGLES

Using only rabbit-ear triangle sink folds, Stacked Triangles boasts a vibrant three-dimensional layering of triangles—with plentiful hexagons a happy by-product.

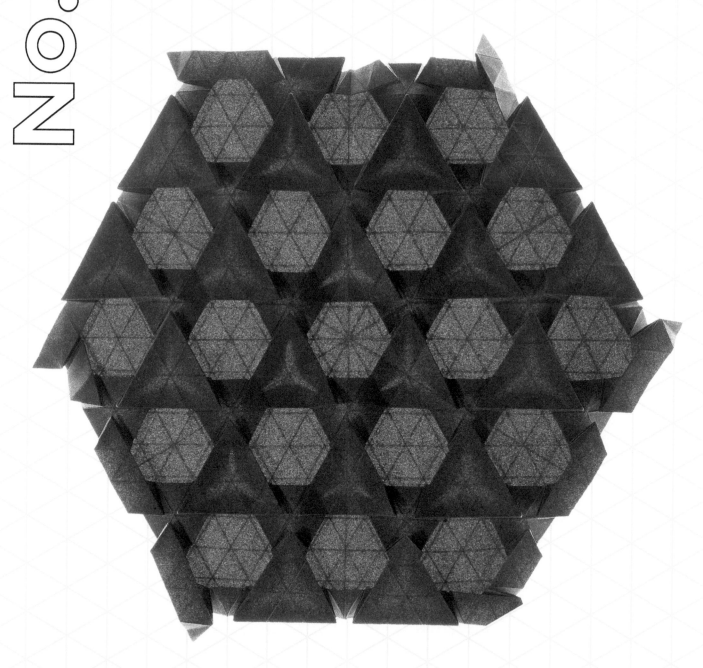

## *techniques*
Triangle Grid, page 7
Rabbit-Ear Triangle Sink, page 14

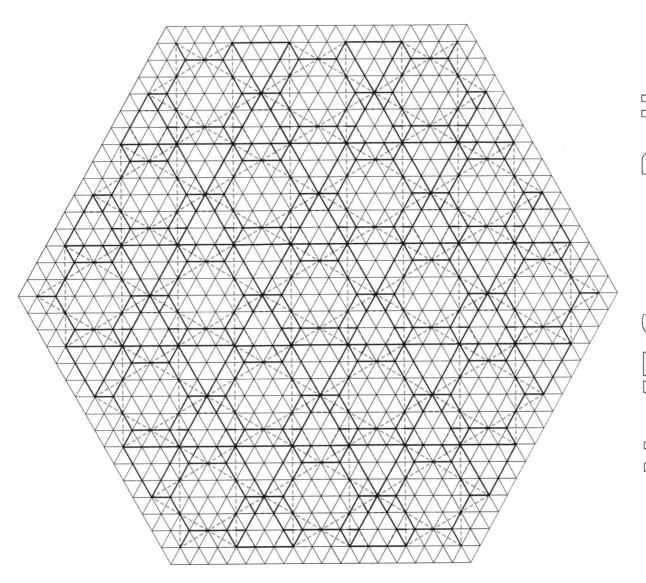

## TIP

Twist folds are like mechanical gears in their rotational patterns. Adjoining twists turn like meshed gears, with each successive twist rotating in the opposite direction.

# Instructions

PHOTO 1

**1** Prepare a triangle grid with the valley-fold precreases shown in the crease pattern. These creases will form all of the piece's rabbit-ear triangle sink folds. Creasing them at the start will save you time and frustration later!

**2** This design is composed of overlapping triangles arranged around hexagonal open spaces. To start folding, locate the central hexagon formed by the valley-fold precreases. Using the hourglass-shaped pairs of offset triangles arranged around the center, fold rabbit-ear triangle sinks to form a triangle that is three grid triangles wide on each side (see photos 1 and 2).

**3** Continue folding these same triangle shapes all around the central hexagonal open space (see photo 3). The triangle tips will overlap one another. Fold them so each triangle is either above or below the next one in succession (see photo 4).

**4** Once you've finished folding the central set of triangles, you have completed one full unit. The pattern just repeats itself to the edge of the paper. Working around the central hexagon, locate each adjacent hexagonal open space on the paper, and repeat steps 2 and 3 for each one.

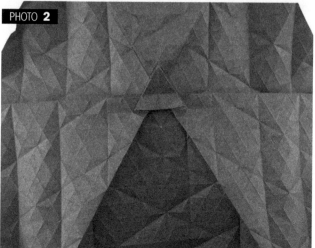

PHOTO 2

PHOTO 3

PHOTO 4

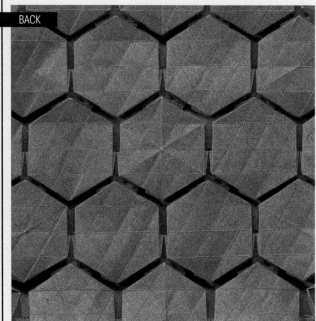

**5** As you fold the triangles, the pattern comes together naturally, and it will be clear where to place the next folds. If you get confused about whether or not a crease is in the correct place, however, check the reverse side of the paper—if all is going well, you will see a pattern of equally sized hexagons (a pattern beginning to take shape in photo 5).

PHOTO **5**

FRONT    BACK

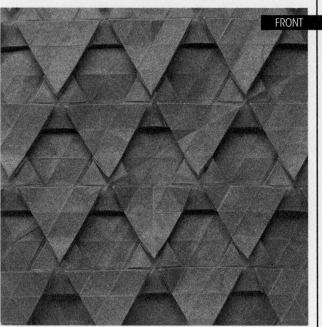

# BIRD BASE

Negative space and positive space create stars and octagons in Christiane Bettens'
eye-popping design. The star shapes in the crease pattern are reminiscent of—
and inspired by—the bird base in traditional origami.

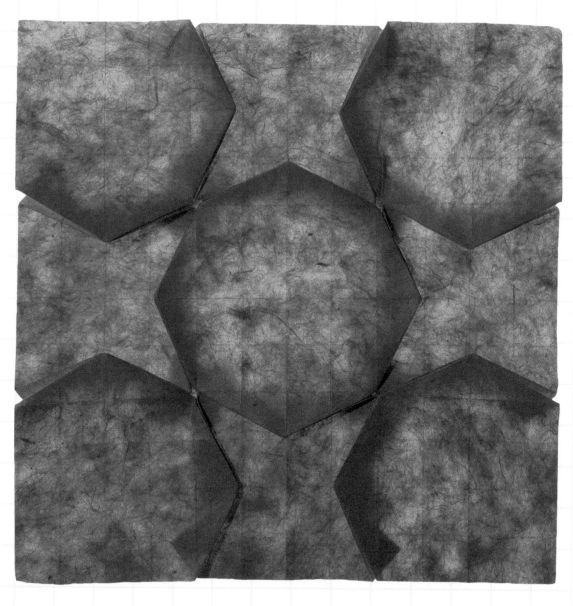

*techniques*
Square Grid, page 6
Square Twist, page 17

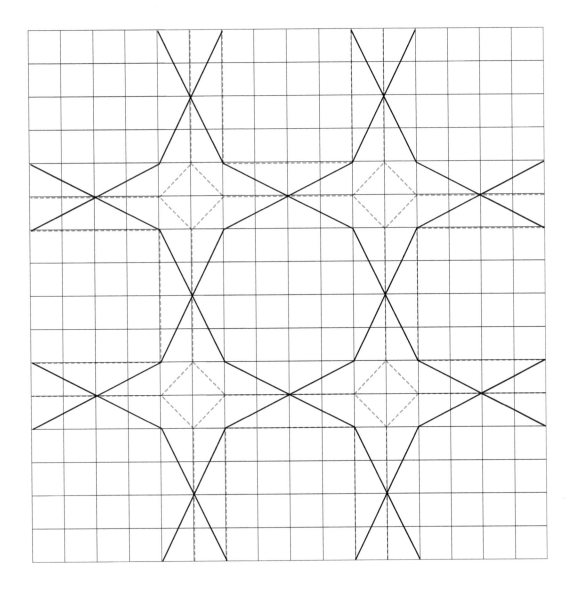

## Instructions

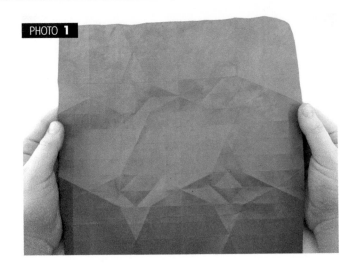

**PHOTO 1**

**1** Using the crease pattern for reference, pinch the mountain and valley folds into place across a 16-by-16 square grid. (For the greatest effect of this magnificent design, I suggest eventually extending the pattern out to fill a square grid of 32 or more divisions.) To more easily make the small-square valley folds, flip the paper over and pinch the folds into place on the reverse side. When your pre-creasing is finished, the paper will resemble a grid of stars connected at the tips, and each star will have a diagonal square in its center (see photo 1 for a front view).

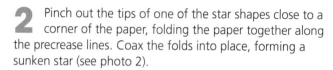

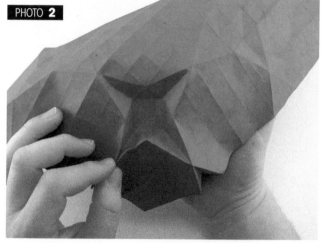

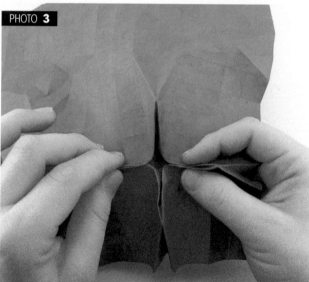

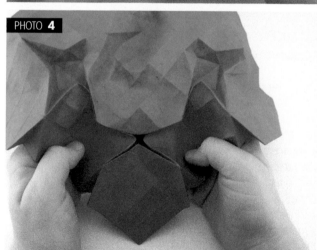

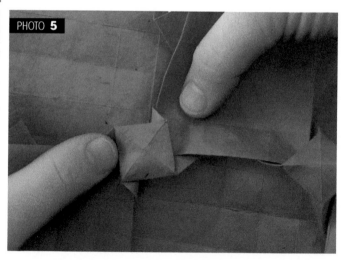

**2** Pinch out the tips of one of the star shapes close to a corner of the paper, folding the paper together along the precrease lines. Coax the folds into place, forming a sunken star (see photo 2).

**3** Bring the inner vertices of the star together, collapsing the star into a plus sign (see photo 3). This step can be tricky—do not force the paper. Be sure to fold along pre-existing crease lines.

**4** Fold the resulting flaps over in the direction indicated by the crease pattern (see photo 4). Fold the flaps of any given star shape either inward or outward—never mix the two orientations. The flaps will become the edges of octagonal tiles.

**5** Repeat steps 2 to 4 across the rest of the paper. You won't be able to fold everything flat just yet, but it's important the paper is folded in the proper directions so everything can fold correctly in the end.

**6** Flip the piece to the back side. You'll see a number of square twists ready to be made. Twist them into shape (see photo 5). Each square twist should have two arms that fold around a pleat and two that fold away from a pleat. Check the alignment of the flaps on the front when making the square twists on the back—the front flaps can drift before you're finished twisting.

**7** When all the twists are complete, flip the paper over to the front and tidy up the front flaps, as necessary.

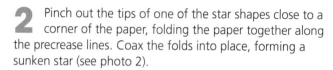

PHOTO **2**

PHOTO **3**

PHOTO **4**

PHOTO **5**

FRONT    BACK

# PROPELLERHEADS

Perceived motion, depth, and curvature—and a sense of play—distinguish the appealing Propellerheads design, which is a cousin of the Water Bomb tessellation.

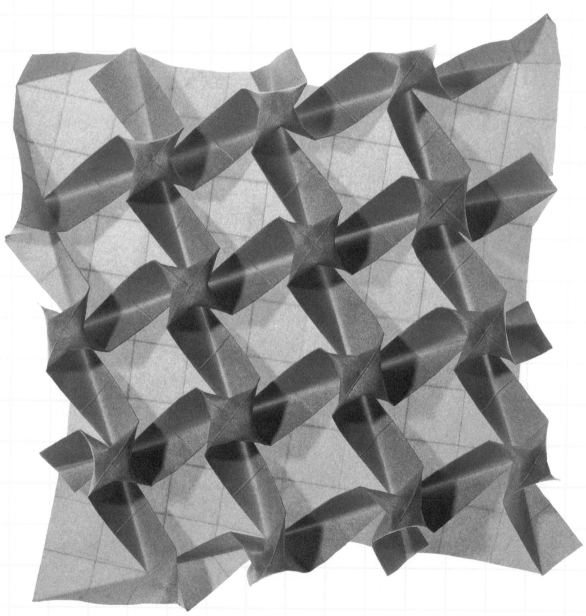

*techniques*
Square Grid, page 6
Square Twist, page 17

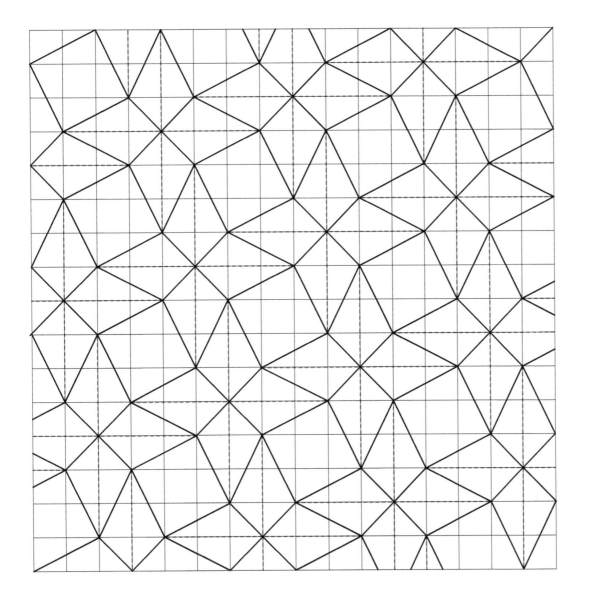

## Instructions

PHOTO **1**

**1** Utilizing a 16-by-16 square grid, precrease all the mountain and valley folds as indicated in the crease pattern.

**2** Flip the piece to the back side. Starting at one of the corners, begin to collapse the folds together along the creases. Don't fold the star-shaped intersections completely; just pinch them together so they start to assume three-dimensional shape, which will prompt adjacent star-shaped intersections to begin to take form (see photo 1).

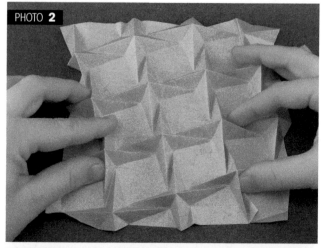

PHOTO 2

3 Having "preshaped" all the star creases, go back over the entire piece, now pinching the star-shaped folds together completely as you move along. You should be able to raise the stars completely at this point (see photo 2).

4 Gently twist the tips of all the stars in the same direction (see photo 3). This step puts a curve into the pleats connecting the stars together and helps lock the design so tension in the paper doesn't pull it apart. Twisting the tips also prepares the paper for the squashing to follow.

5 Grasp two opposing pleats on one of the twisted star tips and pull, but not too much. Repeat with the other two pleats. You are using the same procedure used to squash a square twist, but now you want to flatten each point only partially, so it still retains three-dimensionality (see photo 4).

6 Shape and curve the three-dimensional twists to make the pleats take on a more organic, flowing appearance (see photo 5). For the best results, dampen your fingers slightly, so you moisten the paper as you shape it, helping it take on and retain smoother curves. Don't use too much water, though, or your paper will come apart entirely!

See an extended version of *Propellerheads* on page 119 in the Gallery section.

PHOTO 3

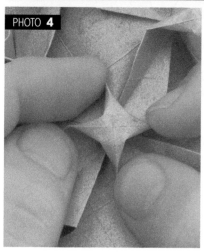

PHOTO 4

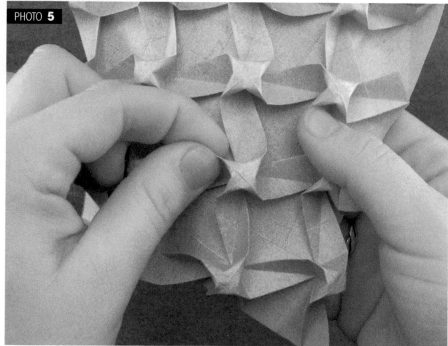

PHOTO 5

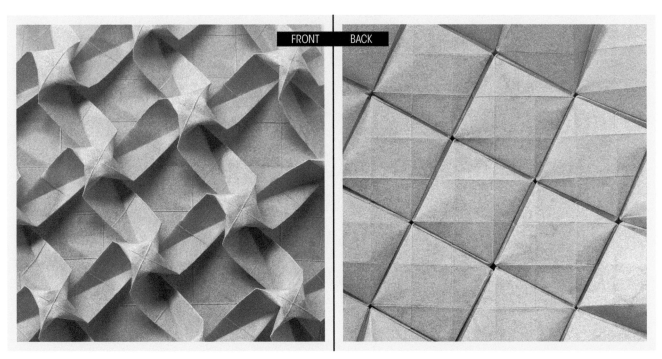

FRONT | BACK

# ARABESQUE FLOURISH

An atypical tessellation constructed from three-dimensional pleat contortions rather than more familiar twists and intersections, Arabesque Flourish is an elegant art piece that's quite easy to fold.

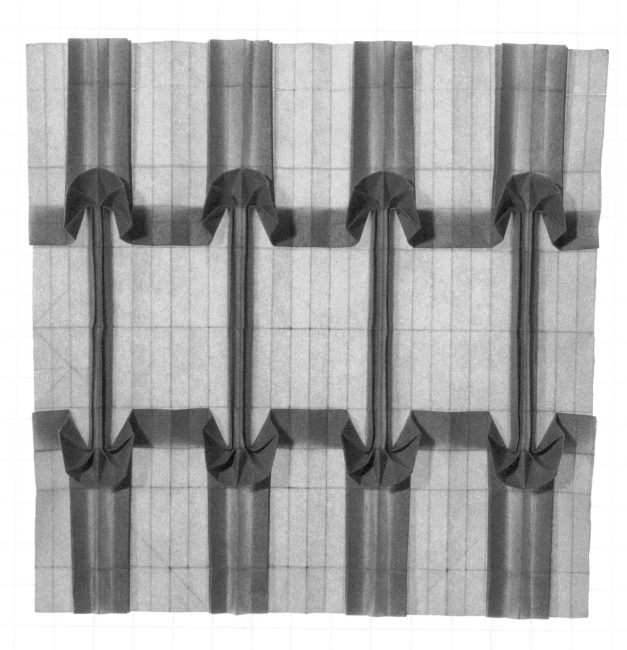

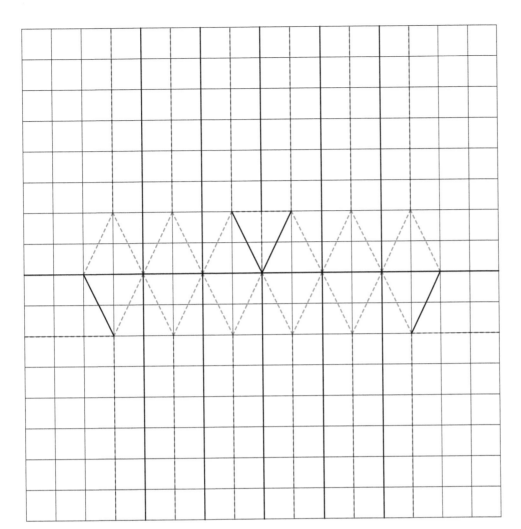

## Instructions

**1** To make one basic Arabesque Flourish unit, divide a square sheet of paper into 16 pleats. Turn the paper 90 degrees and fold the paper into fourths. Fold both of the two middle wide pleats in half, so four narrower pleats are in the center with a wide pleat at each end of the paper. Use the crease pattern for reference.

**2** Again referring to the crease pattern, fold the diagonal lines connecting the middle crease line—the one that runs across all the pleats in the center of the paper—with the parallel lines directly above and below it. Pinch the paper together at both end points, and squeeze the creases into place. The diagonal lines will run on both sides of the central crease line in a pair of zigzag patterns.

**3** Once you've completed the precreasing, fold the pleats together accordion-style. Fold the central

horizontal pleat over, essentially folding the paper in half. You are beginning the process of curving the flourish element into shape (see photo 1).

PHOTO **1**

PHOTO **2**

**4** Pull the pleats together, causing the central horizontal crease line to curve. Fold the pleat formed by the central horizontal crease line and its parallel creases toward you on the left-hand side—it will fold flat along the edges of the paper and curve the central crease line (see photo 2). Do the same on the right side of the paper.

**5** Above the flourish element, fold the two leftmost pleats over to the left, on top of one another. Open the central pleat and flatten the paper, so the pleat is completely flat (see photo 3). Mirroring the left side, fold the two rightmost pleats over to the right, and repeat the flattening process.

**6** Still working above the flourish, squeeze the central vertical pleat back into a vertical position. This move brings the entire design together, allowing the five pleats at the bottom of the flourish to stand upright (see photo 4).

**7** All that remains for you to do is some shaping of the fan-shaped pleats on the flourish itself. Fold them over a bit and press tightly with your fingers to mold them into a more curved, organic shape.

For best results with this tessellation, use relatively heavy paper. During the final shaping, dampen your fingers or the paper itself very lightly, and then hold the desired shape until the paper dries.

Note: These instructions and instructional photographs explain how to make a basic Arabesque Flourish unit. To make the larger design, fold a larger piece of paper with more pleats. Because the pleats extending from the base of the flourish are all packed together, turn the paper 180 degrees and use the same pleats to fold a mirror image of the pattern, so the two sides match up.

PHOTO **3**

## TIP

Backlighting tessellations is one of the best ways to display them—and also gives great insight into how they are constructed. Many origamists design tessellations specifically for this style of display.

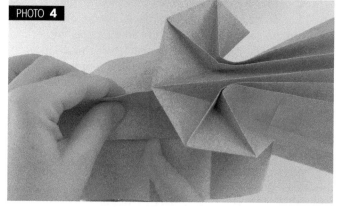

PHOTO **4**

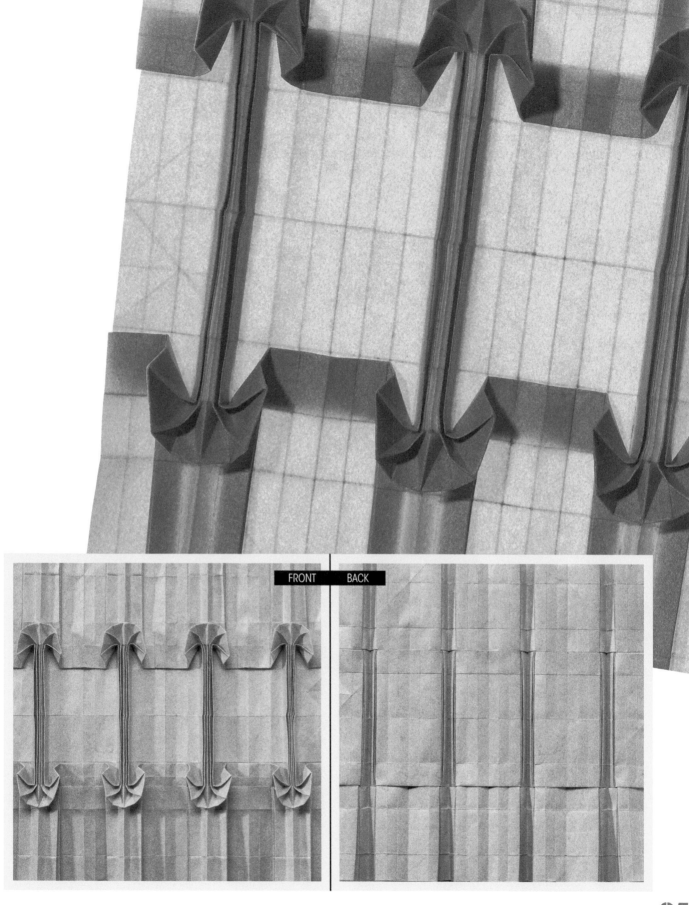

FRONT    BACK

# FIELD OF STARS

Combining twist, pleat, and sink techniques, this tessellation creates a variety of shapes in both positive and negative spaces. The resulting pattern is complex, sprawling, and enchanting.

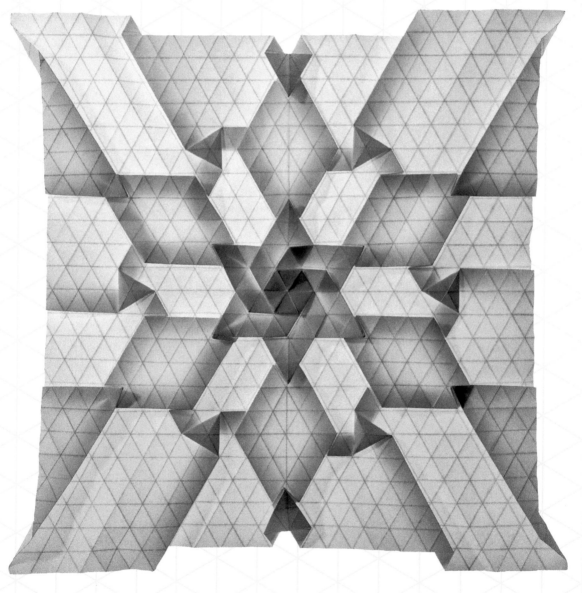

## techniques

Triangle Grid, page 7
120-Degree Pleat Intersection, page 9
Rabbit-Ear Triangle Sink, page 14
Triangle Twist, page 16
Hexagon Twist, page 18

## No.21 Crease Pattern

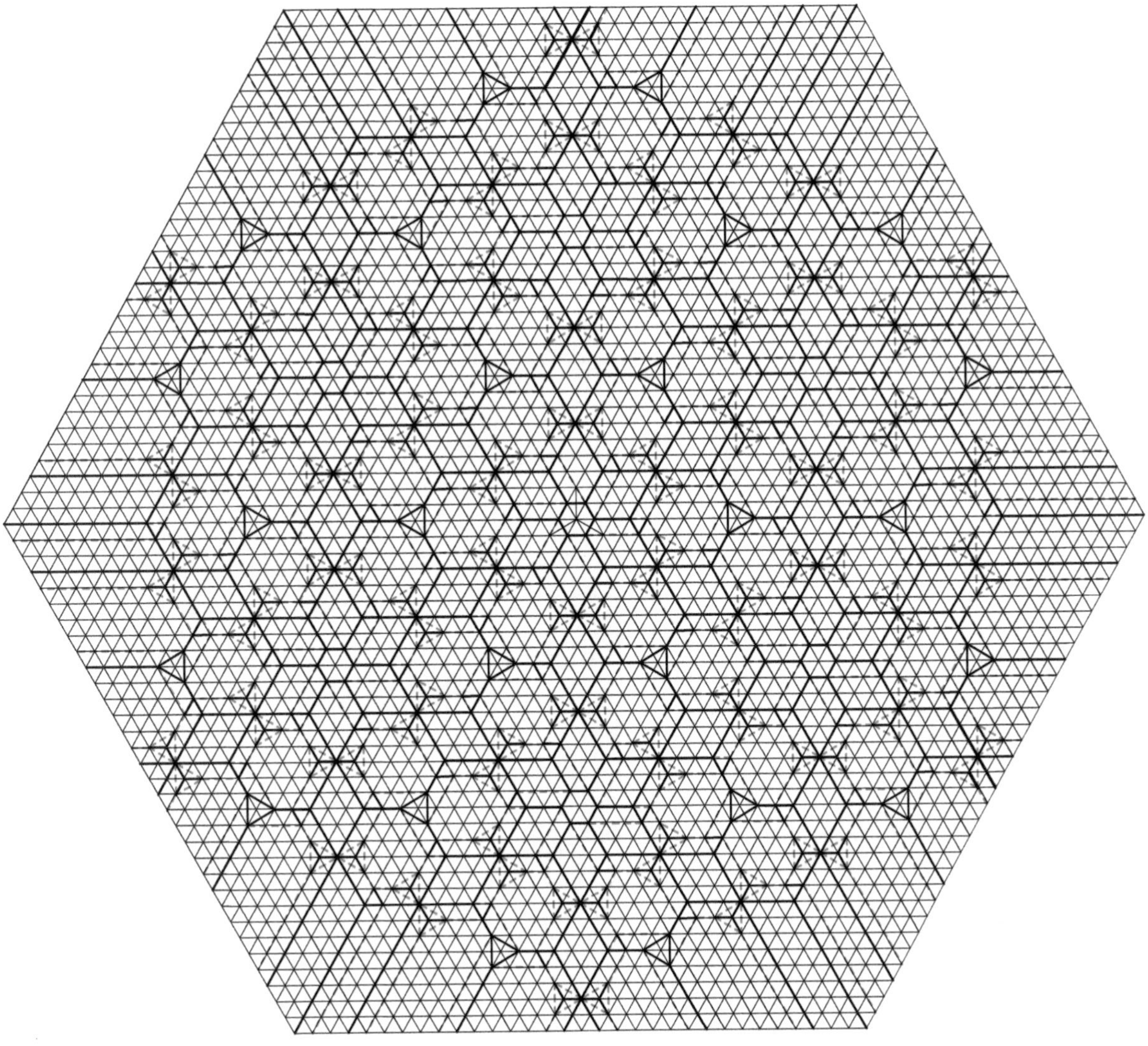

## *Instructions*

PHOTO 1

PHOTO 2

**1** Precrease a triangle grid with at least 64 divisions. Identify the center of the grid and fold a hexagon twist.

**2** Fold a star with petal shapes exactly like you did for the Star Twist tessellation on page 67, but instead of folding a 60-degree pleat intersection at the points of the star, fold a tip using the rabbit-ear triangle sink (see photo 1).

**3** Repeat step 2 around the central hexagon twist until you have a full star (see photo 2). You can make this step easier by precreasing the triangle-sink folds before folding them, but be careful—placement of the triangle sinks can be tricky. Refer to the crease pattern for specific crease locations.

**4** Fold a 120-degree pleat intersection on a pleat extending from the tip of the star, and then fold a matching intersection on the pleat adjacent to it. The intersecting pleats in the center form two legs of a triangle; fold a third pleat so it meets the first two, and create a triangle twist (see photo 3). Pay close attention to the direction in which you twist it—all the triangle twists on the piece should match the rotational direction of this first one.

**5** Create another triangle twist on the next set of pleats radiating from the center star. You now have created a diamond shape between the triangle twists, which will connect the central star with the next star as the pattern radiates out. Using a rabbit-ear triangle sink, fold a tip at the end of the diamond shape (see photo 4). This tip is the first tip of the next star twist.

**6** Repeat step 5 all around the center star, making triangle twists and diamond shapes. Use the outer tip of each diamond shape as a reference point for the creation of the next star twist.

**7** Finish the edges by folding the excess paper and extending the pleats in a clean, pleasing manner that complements the central design (see photo 5).

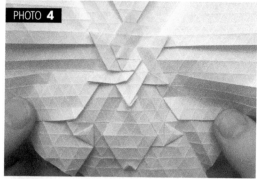
PHOTO **4**

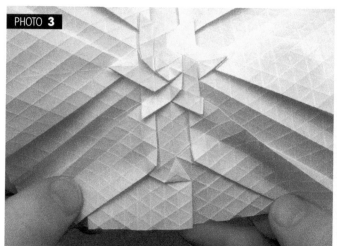
PHOTO **3**

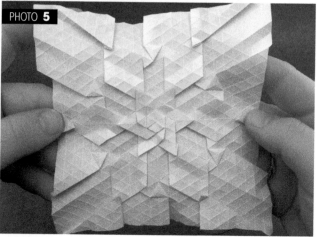
PHOTO **5**

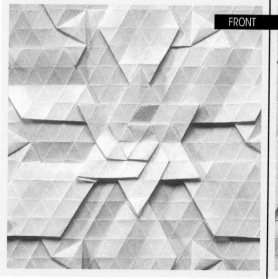

FRONT    BACK

# NEGATIVE SPACE STARS

Negative space defines a central star cut out of a larger surrounding star in a stunning pattern. The twist is there are no twists in this star-centric piece! In many ways, this tessellation is less about what's in it and more about what's missing.

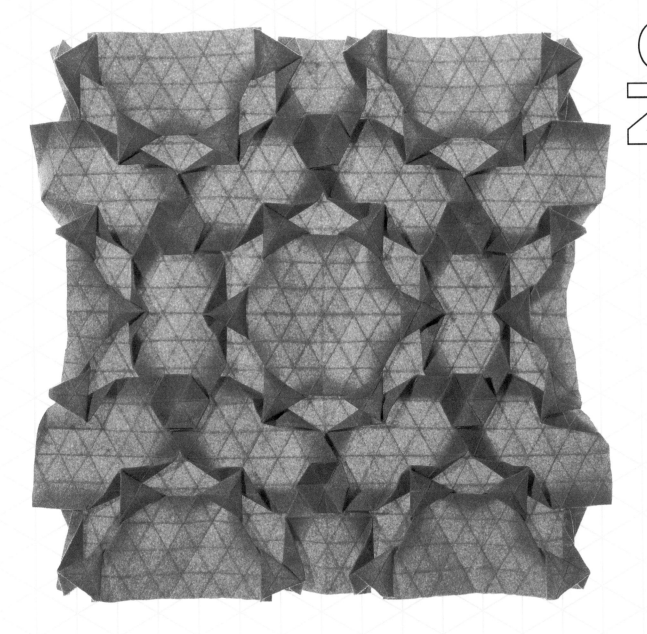

*techniques*
Triangle Grid, page 7
60-Degree Pleat Intersection, page 12

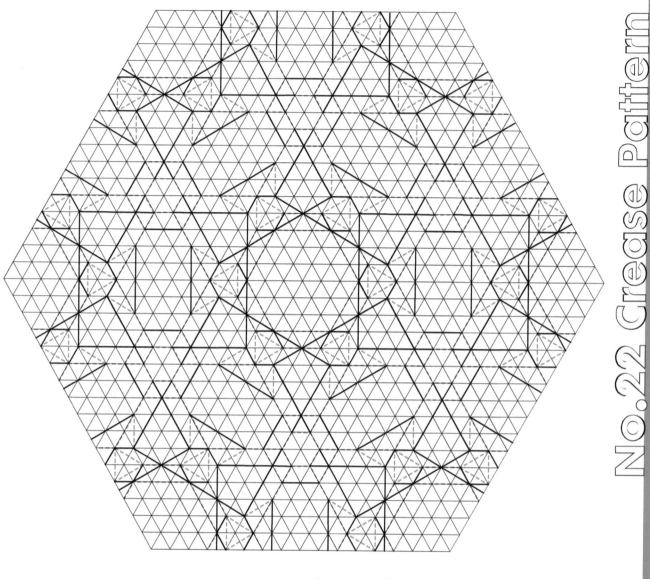

## Instructions

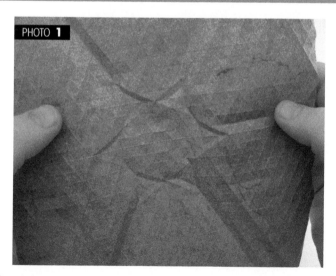

**1** Precrease a triangle grid folded into at least 48 divisions. Using a hexagon-shaped grid will enhance the symmetry of the finished piece.

**2** Locate the center of the grid. Utilizing the center point as your reference point, precrease the offset mountain and valley folds (see photo 1), which will make folding the negative star much easier than it would be folding the creases as you go along.

**3** Fold the lines you creased in step 2, collapsing the triangular shapes together. Initially, fold them just enough to allow the next one to collapse. Once you've partially collapsed all six triangles, work back around and fold them all more tightly. You will create pleats extending straight out from the corners of each triangle (see photo 2).

**4** By collapsing the triangular shapes in step 3, you created small pleat flaps around the central open area. Fold these pointed little pleat shapes toward the center, being careful to keep the pleats tightly folded so the pattern doesn't drift apart (see photo 3).

**5** Choose one of the sunken triangles, and invert the pleat extending outward from it. This move creates a bit of an awkward shape, but it allows you to fold two pleats extending away from each other at a 120-degree angle (see photo 4). These two pleats meet at the tip of the triangle on the back side—actually forming an inverted triangle twist!

**6** Flip the piece over to the other side, and fold the two pleats toward the center of the star. To do this, you have to fold a bit of paper backward at the tip of the triangle, so the tip of the triangle twist sits on top of the folded paper (see photo 5). Refer to the crease pattern for specific folding locations. The two pleats form the sides of one of the kite shapes that will make up the outline of the larger star.

PHOTO **2**

PHOTO **3**

PHOTO **4**

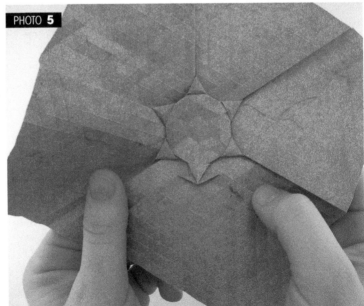

PHOTO **5**

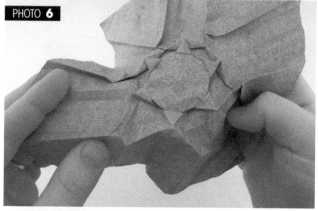

PHOTO **6**

**7** Repeat steps 5 and 6 for the remaining five sides of the star. Fold the pleat intersections at the tips of the larger star with 60-degree pleat intersections (see photo 6). These folds will become part of a hexagon twist after you've folded all the adjoining star units, similar to the method used for the Star Twist tessellation on page 67.

**8** When you've folded all six sides of the star, you've completed the main tessellating unit of Negative Space Star (see photo 7). Tile units together to make a tightly packed design, as in the crease pattern, or use them to create your own complex designs.

See an extended version of *Negative Space Stars* on page 119 in the Gallery section.

PHOTO **7**

## TIP

"Floating" glass frames provide wonderful display venues for tessellations. Hang your framed pieces in front of a window that gets bright sunlight.

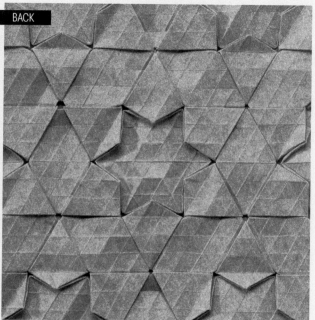

FRONT    BACK

# 3.4.6.4

This beautiful, complex tessellation uses 3.4.6.4 geometry to create a pattern of hexagons, squares, and triangles. In theory, the 3.4.6.4 tiling shouldn't fit on a grid—at least not without heavy modification. So, this pattern takes a few shortcuts, including a slightly non-squarish square twist that's a little tricky to fold.

*techniques*
Triangle Grid, page 7
Triangle Twist, page 16
Square Twist, page 17
Hexagon Twist, page 18

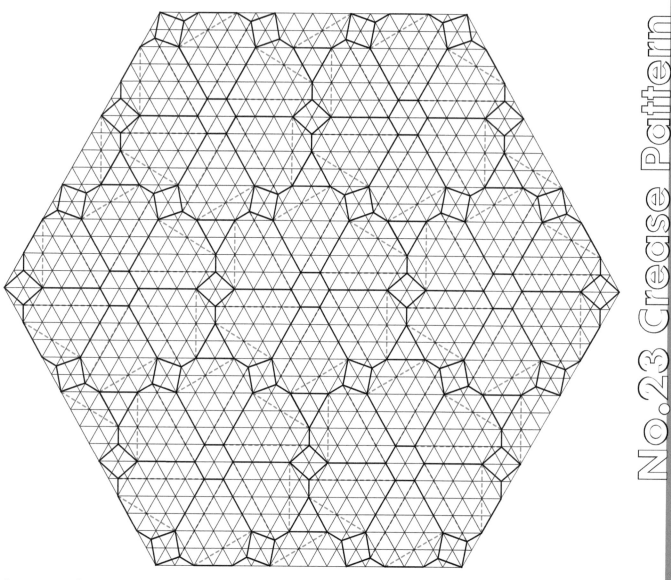

## Instructions

**1** Precrease a triangle grid with at least 32 divisions.

**2** Fold all the offset precreases indicated on the crease pattern. These offset creases form an additional set of pleats from which you'll create square and triangle twists. In this project you essentially lay a tessellation tiling of all triangle twists on top of a tiling of all hexagon twists. And the intersections between the two create the square twists!

**3** Fold a hexagon twist in the center of the grid, in the middle of the larger hexagon made by the offset creases.

PHOTO **1**

**4** Following one of the pleats extending out from the central hexagon twist, find the point where it intersects an offset pleat (see photo 1). Using the existing pleat as a guide, fold the offset pleat in the same rotational direction to create a square twist (see photo 2).

**5** The offset pleats form a triangle twist point. Fold a triangle twist from these pleats, using the rotational direction of the square twist as your guide. Keep moving in the same direction around the central hexagon, and fold the next square twist adjoining the triangle twist (see photo 3). Continue folding as you move around the central hexagon twist, forming square twists and triangle twists one after another until you complete the circle (see photo 4).

See an amazing design using 3.4.6.4 symmetry by Chris Palmer on page 116 in the Gallery section.

PHOTO **2**

PHOTO **3**

PHOTO **4**

FRONT    BACK

# AZTEC TWIST

A synthesis of two other tessellations—Stacked Triangles and Tiled Hexagons—this design derives a rich complexity from the interplay between two patterns that prove greater than the sum of their parts.

*techniques*
Triangle Grid, page 7
Rabbit-Ear Triangle Sink, page 14
Triangle Twist, page 16
Hexagon Twist, page 18

*Instructions*

PHOTO **1**

**1** Using a triangle grid and the crease pattern as your guide, create a pattern of triangles overlapping each other, similar to the Stacked Triangles folds on page 80 but making each triangle six grid triangles long on each side (see photo 1).

**2** Unfold the stacked triangles and precrease all the offset creases, which will form a pattern of tiled diamond shapes on top of the existing design (see photo 2). These shapes will become the tessellation's triangle and hexagon twists.

**3** Fold a central hexagon twist and the surrounding triangle twists. These twists will not match up with the precreased grid, so twist and squash them into place carefully (see photo 3). The pleats extending from the triangle twists toward the paper's edges won't fold flat properly. Don't worry: You'll fold them into their final resting place shortly.

**4** Fold the original triangles (from step 1) back into the paper (see photo 4). This step can be difficult, because the pleats extending from the center interfere with the triangle-sink folds at the tips of the triangles. You have to fold one offset pleat—the one folded over in the "wrong" direction to keep the two offset pleats from making a point—to face the opposite direction. This fold allows the two offset pleats to meet, so you can fold a rabbit-ear triangle sink that allows the tip to fold flat. Leave the other offset pleat as is.

**5** Work your way around the paper, folding the original triangle folds into place and rearranging the offset pleats to create additional folded points around the edge of the pattern (see photo 5).

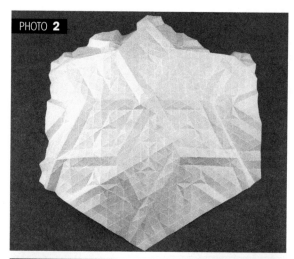
PHOTO **2**

PHOTO **3**

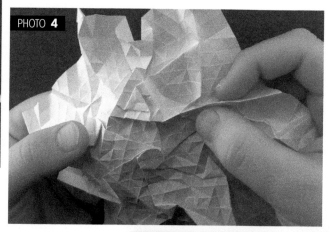
PHOTO **4**

PHOTO **5**

## TIP

A light bleach solution, carefully and sparingly sprayed on heavier papers, will bleach interesting patterns along origami crease lines. Experiment with applying bleach before you start folding and after you finish.

**6** For a final touch, you can fold the offset pleats in the center in the opposite direction. To do this, you must retwist the central hexagon twist. This pleat reorientation allows the piece to fold flat more easily and adds some visual interest to the center portion of the design (see photo 6).

PHOTO **6**

FRONT    BACK

# No.25 ARMS OF SHIVA

Intricate and dazzling, Arms of Shiva looks fantastic from either side of the paper. This design introduces a new shape, too: a teardrop twist made with a stretched pentagon.

## *techniques*

## Instructions

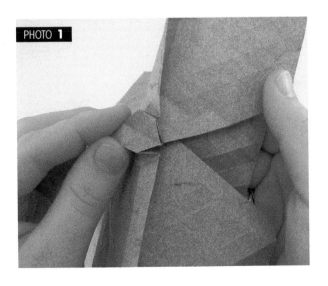

PHOTO **1**

**1** Precrease a triangle grid with 64 or more divisions—the more folding space available, the better. Use a light paper, because this tessellation is wonderful when backlit.

**2** Precrease the offset folds shown on the crease pattern. Folding them precisely will make it much easier for you to collapse them later, when they become pleats and a large hexagon twist.

**3** Fold a teardrop twist, using the crease pattern as a reference. Line up the twist on the central offset crease, with the teardrop pointing toward the center. Fold the back end of the teardrop like a 60-degree pleat intersection, overlapping the layers. Fold the next two pleats extending out from the sides of the teardrop away from the center of the design, so they both face outward (see photo 1).

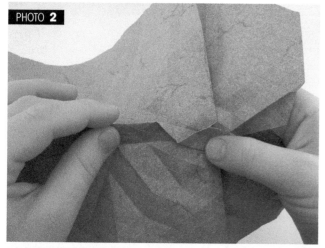

PHOTO 2

**4** Align the pointed tip of the teardrop using the large pleat created by the offset creases. Fold this large pleat over, and then fold the tip of the teardrop back over so the tip lines up with the center of the paper. This move creates a zigzag in the pleat, but it allows the teardrop twist to lie flat (see photo 2).

**5** Follow one of the pleats extending out from the side of the teardrop. Three pleat widths away, fold a 120-degree intersection. Extend both inside pleats outward. The new inside pleat will connect to the side of another teardrop twist; use it as a reference to fold another teardrop twist into shape (see photo 3).

**6** Continue around in circular fashion to fold six teardrops in total. You'll encounter a large amount of paper buildup in the center due to all the large pleats. Using the offset precreases, make a large hexagon twist in the center. All the large pleats should lie flat at this point (see photo 4).

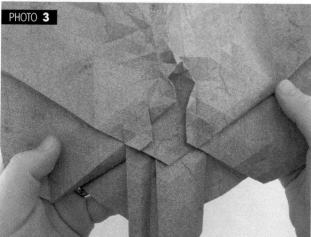

PHOTO 3

PHOTO 4

## TIP

Many papers can be painted with mineral oil to turn them into sheer, translucent origami tessellations. Try this maneuver only after your projects are fully completed!

**7** At each corner of the large hexagonal unit traced out by the far edges of the teardrops, fold additional hexagon twists to tessellate the pattern. Along one of the single pleats extending out from the corners of this large, underlying hexagon, fold a hexagon twist centered on a point four pleat widths from the tip of the 120-degree intersection (see photo 5). Use the crease pattern to determine the exact placement of the fold.

**8** Work your way around the design, folding hexagon twists at each corner. All these twists should rotate in the same direction.

**9** Locate where the pleats extending from the hexagon twists intersect with the pleats coming out of the top of the teardrop twists, and fold a 120-degree pleat intersection. The two pleats that extend from these two intersections right above the teardrop twist will meet and form the top of another teardrop twist, as part of an adjacent large hexagonal unit (see photo 6). The pattern tiles infinitely and always will connect in the same way.

See an extended version of the *Arms of Shiva* tessellation on page 118 in the Gallery section.

PHOTO **5**

PHOTO **6**

FRONT    BACK

# Gallery

**Robert J. Lang**
*Stars and Stripes, opus 500*, 2007, Wyndstone Marble paper

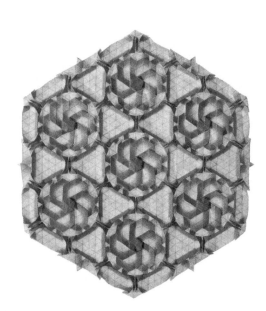

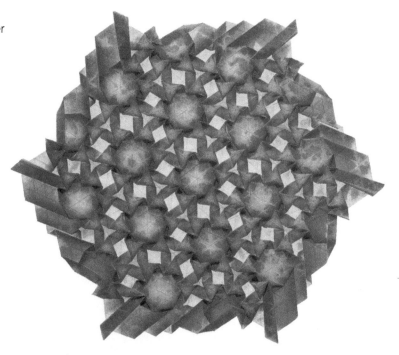

**Christine Edison**
*Roundabout*, 2006, Elephant Hide paper

**Joel Cooper**
*Snowflake Weave*, 2006, bleached Elephant Hide paper

**Christiane Bettens**
*Ryan's Tessellation,* 2006, Elephant Hide paper

**Joel Cooper**
*Tessellated Fujimoto Hydrangeas,* 2005, lokta paper

**Joel Cooper**
*Mask,* 2006, Elephant Hide paper

**Christine Edison**
*Modern Blue,* 2006, kami paper

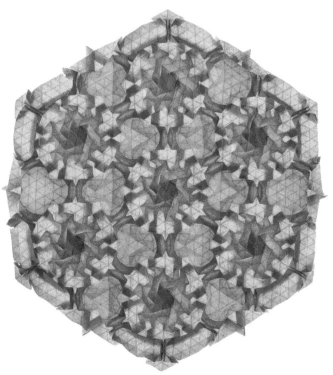

**Christine Edison**
*Autumn Leaves,* 2006, Elephant Hide paper

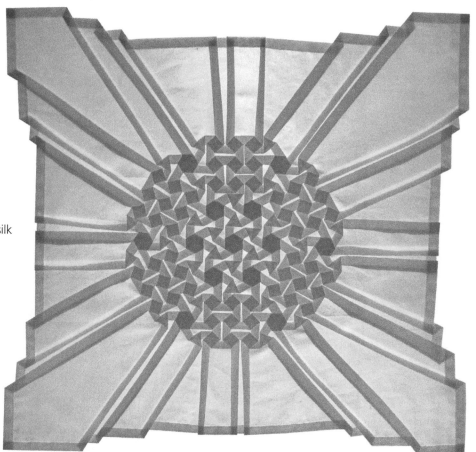

**Chris K. Palmer**
*Watering Fujimoto's Garden,* 1995, silk

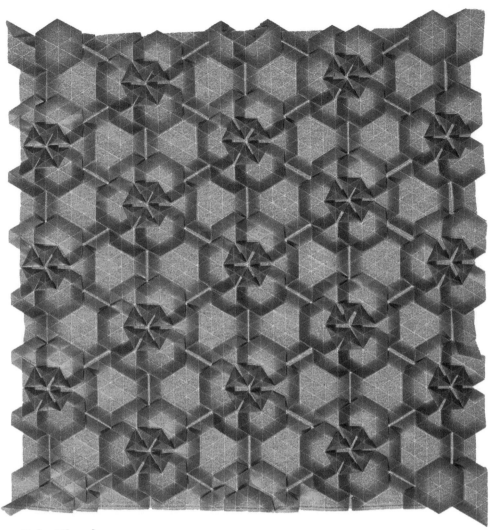

**Eric Gjerde**
*Flowering Grid,* 2006, metallic kami paper

**Polly Verity**
*T-Pentomino (pared),* 2007, polypropylene

**Sipho Mabona**
*Fugu,* 2006, Origamido paper

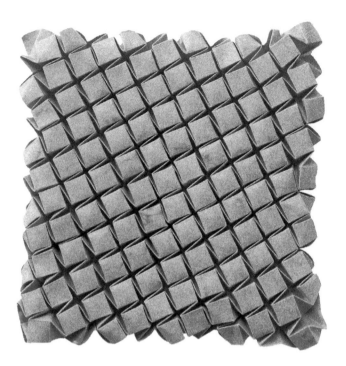

**Eric Gjerde**
*Water Bomb* (extended version), 2006, Elephant Hide paper

**Eric Gjerde**
*Star Twist* (extended version), 2005, Elephant Hide paper

**Eric Gjerde**
*Arms of Shiva* (extended version),
2006, wood-pulp and linen paper

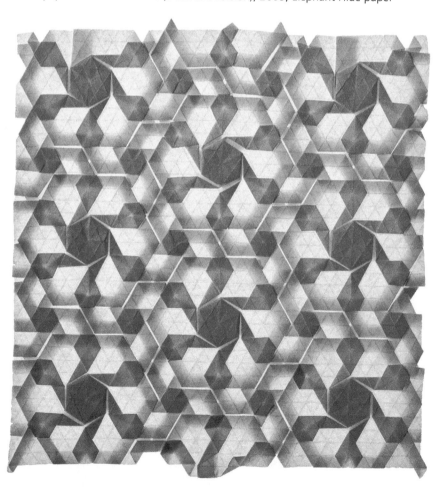

**Eric Gjerde**
*Negative Space Stars* (extended version),
2006, glassine

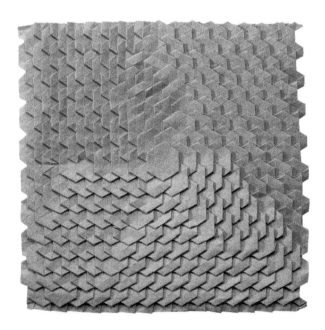

**Eric Gjerde**
*Spread Hexagons* (extended version),
2005, recycled-wood-pulp paper

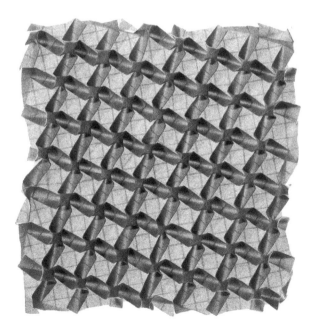

**Eric Gjerde**
*Propellerheads* (extended version),
2006, Elephant Hide paper

# Acknowledgments

Thanks to all the members of the origami community who helped me in the creation of this book. Special thanks go to Jane Araújo, Alex Bateman, Christiane Bettens, Joel Cooper, Christine Edison, Ralf Konrad, Sipho Mabona, Miguel Angel Blanco Muñoz, and Polly Verity for their contributions and assistance with designs. Philip Chapman-Bell was instrumental in helping me make sense of complicated material and provided support. Robert J. Lang and Chris Palmer supplied artistic inspiration in new directions for origami tessellations.

Thanks go to Dr. Keith Critchlow for his book, *Islamic Patterns: An Analytical and Cosmological Approach* and the late Owen Jones for his classic tome *The Grammar of Ornament*. Both of these books provided endless ideas and insight into geometric patterns and design.

My family and friends deserve a special mention for coping with a paper-obsessed maniac while preparing the material for this book. They never once told me to "put the paper away."

Thanks to Robert J. Lang proposing a good book idea for A K Peters, Ltd.

A special thank you to Jeffrey Rutzky for his design and production expertise, as well as his determination that I get this book published. His help has been invaluable.

Most of all, I wish to thank Bekah Gjerde. I could not have done this book without her unwavering support and assistance. Her patience for my constant folding is simply the stuff of legend.

## PHOTOGRAPHY CREDITS

Photos copyright Brian McMorrow, 1999–2007: pages 5, 18, 20, Jameh Mosque; pages 6, 13, Madraseh-ye Chahar Bagh; page 9, dome of the central pavilion of the tomb of Hafez; page 14, Regent's Mosque.

Photo copyright Dwight Eschliman, 2007, page 114, *Stars and Stripes, Opus 500*.

# About the Author

When asked by his parents what he wanted to be when he grew up, five-year-old Eric replied, "paperologist." Throughout his childhood and adolescence he enjoyed paper crafts and origami—frequent birthday gifts were stacks of paper and rolls of tape.

After preparing for a technology career, Eric kept looking for an artistic outlet to balance his creative side with his professional life. He returned to his childhood love of paper, and discovered a deeper appreciation for the beauty of origami by exploring and researching new areas of folding. These days, Eric focuses on the geometric art of origami tessellations.

Eric continues to fold, teach, and share his talent through art workshops, origami conventions, international art exhibitions, and through his popular site on the Internet. He currently balances dual careers as a technology professional and paper artist in Minneapolis. You can learn more about Eric's passion for paper art at origamitessellations.com.

# Index

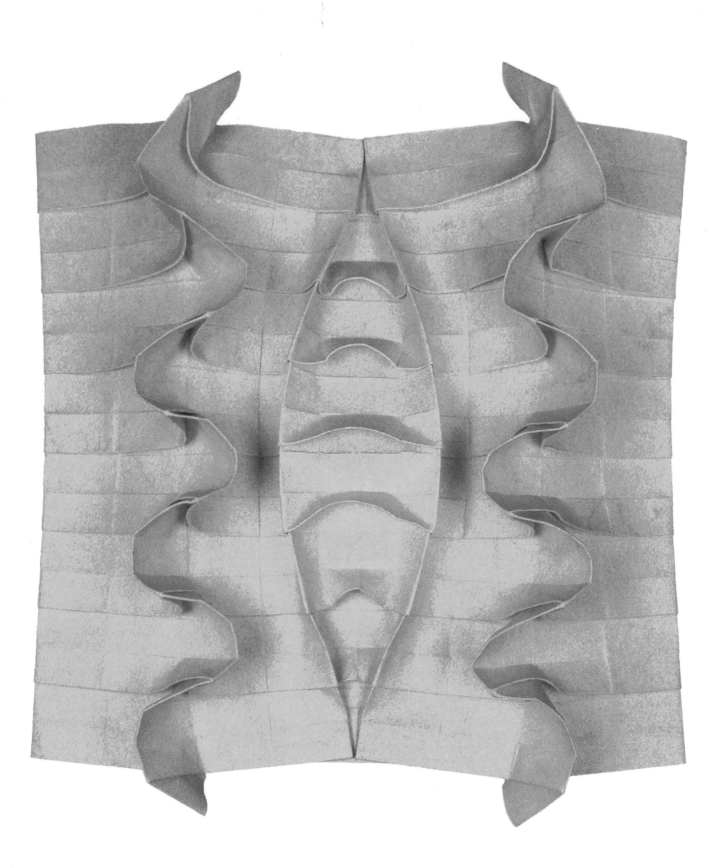